Managing Pharmaceuticals in International Health

Stuart Anderson
with Reinhard Huss, Rob Summers
and Karin Wiedenmayer

Foreword by Richard Laing

Birkhäuser Verlag
Basel · Boston · Berlin

Authors

Dr. Stuart C. Anderson
Department of Public Health and Policy
London School of Hygiene and Tropical Medicine
Keppel Street
WC1E 7HT London
United Kingdom

Dr. Karin A. Wiedenmayer
Swiss Centre for International Health
Swiss Tropical Institute
Socinstrasse 57
4002 Basel
Switzerland

Dr. Reinhard Huss
Department of Tropical Hygiene and Public Health
Ruprecht-Karls-University of Heidelberg
Im Neuenheimer Feld 324
69120 Heidelberg
Germany

Prof. Robert S. Summers
School of Pharmacy
Medical University of Southern Africa (MEDUNSA)
P. O. Box 218
MEDUNSA 0204
Republic of South Africa

Library of Congress Cataloging-in-Publication Data

Managing pharmaceuticals in international health / Stuart Anderson, with Reinhard Huss
    ... [et al.] ; foreword by Richard Laing
      p. cm.
    Includes bibliographical references and index.
    ISBN 3-7643-6601-X (alk. paper)
    1. Pharmaceutical policy. 2. World health. I. Anderson, Stuart (1946-

    RA401.A1M364 2004
    362.17'82--dc22   2004046884

Bibliographic information published by Die Deutsche Bibliothek
Die Deutsche Bibliothek lists this publication in the Deutsche Nationalbibliografie;
detailed bibliographic data is available in the Internet at <http://dnb.ddb.de>.

ISBN 3-7643-6601-X Birkhäuser Verlag, Basel – Boston – Berlin

© 2004 Birkhäuser Verlag, P.O. Box 133, CH-4010 Basel, Switzerland
Part of Springer Science+Business Media
Printed on acid-free paper produced from chlorine-free pulp. TCF ∞
Cover design: Micha Lotrovsky, CH-4106 Therwil, Switzerland
Cover illustration: The Life Cycle of Medicines (see page 7)
Printed in Germany
ISBN 3-7643-6601-X

9 8 7 6 5 4 3 2 1   www.birkhauser.ch

Contents

Foreword

"One third of the world's population lack effective access to quality assured essential medicines used rationally".

When WHO first made this statement fifteen years ago, there was general concern that medical miracles such as antibiotics, antiparasitic medicines, vaccines and analgesics would not be available to many people. Today, the proportion of those lacking access is lower in Asia and Latin America and higher in Africa but there are probably about two billion people in this situation. This book describes the many problems involved, and then puts together possible solutions based on country experiences in a comprehensive and coherent manner.

Many people lack access to essential medicines because they and their countries are poor, and because of inefficiencies in their health systems. We know that in low and middle income countries between 25 and 40 per cent of health expenditure is on medicines, and that most of that expenditure is out of pocket. Often this amounts to less than US $ 2 per head per year! In contrast, high income countries spend only 8 to 15 per cent of health expenditure on medicines, and this is mostly paid for by health insurance or social security funds. High income country expenditure may be over US $ 400 per person per year! So managing the scanty resources available in low income countries becomes all the more important. Inefficiencies are becoming an increasing financial burden for many people even in high income countries; but the poor in low income countries may pay for any waste or managerial failures with their health or life.

When we look at what poor people spend their precious resources on, we discover many related problems needing solutions. The quality of commonly used antimalarial and TB drugs have recently been shown to be substandard in some African and Asian countries. Prices of widely used brand and less commonly used generic products may vary one hundredfold. Taxes, duties and excessive markups may more than double the prices paid by consumers. We see irrational use of injections, with more than 15 billion injections given per year, and antibiotics being frequently prescribed unnecessarily. Finally, we have very little research for new medicines for neglected diseases such as malaria, TB, trypanosomiasis and leishmaniasis.

But for all of these problems we see solutions existing in different environments. WHO has led the way with selection by providing a Model Essential Medicines List. In addition, the WHO Medicines Library provides all the needed information to select national Essential Medicines Lists. Pool procurement efforts in regions as different as those of the Eastern Caribbean or the Persian Gulf have enabled countries that have combined together to obtain quality assured essential medicines at very competitive prices. WHO in conjunction with other UN organizations now provides a prequalification service to identify the AIDS, TB and malaria products of assured quality. Price information is more widely available than ever before through the efforts of consulting companies, Médecins sans Frontières (MSF) and the WHO regional office in Africa.

The rational use of medicines has improved in some countries for several reasons. A dramatic drop in the use of injections in Indonesia, the effective treatment of pneumonia in children, and of sexually transmitted infections (STIs) in adults according to guidelines, are all signs of improvement. AIDS is being treated according to modern treatment guidelines in Brazil and other low and middle income countries. Public Private Partnerships such as the Medicines for Malaria Venture (MMV) or Drugs for Neglected Diseases (DNDI) have been established to ensure that research is undertaken to discover the new medicines for neglected diseases. The European Union is providing support for a clinical trials network to test these new medicines in the environments where the diseases occur. Finally, there has been unprecedented international support for the Global Fund to fight AIDS, tuberculosis and malaria. A major portion of these resources will go towards the purchase, distribution and measures to ensure the rational use of these medicines.

This book has been written by experienced teachers and practitioners to bring together personal and global experiences in an accessible fashion. It will be useful for international public health practitioners who need to know about pharmaceuticals, and pharmacists who need to know about international public health. Additionally, activists, NGO members, donors and UN officials may find the information useful for their work.

When poor people have limited resources and great needs, their lives depend on these scanty resources being spent as efficiently as possible. Managing pharmaceuticals in international health in a way which provides the greatest possible benefit for those in greatest need should be the goal of any public health practitioner or pharmacist working in the field. This book provides the information needed to understand the real causes of many access problems, and suggests possible solutions to them. The challenge now is to take these lessons and implement them, to ensure universal access to essential medicines.

Richard Laing
Geneva, March 2004

Preface

The ready availability of essential medicines to entire populations is a key determinant of public health in all countries. Even in the remotest parts of the world people consume modern medicines. Yet the 75 per cent of the world's population living in low and middle income countries consume barely 20 per cent of the world's medicines. And just eight low and middle income countries consume over 60 per cent of all the medicines available in the developing world.

Addressing these great inequalities is one of the major challenges in international health. But the issues surrounding medicine use are complex. Many of the problems can only be tackled at a global level by many agencies, governments and pharmaceutical companies working together. An enormous literature now exists that addresses the economic, social and political factors involved.

Despite this, pharmaceutical policy issues receive only scant attention in textbooks of international health. For the student of international health the task of getting to grips with the subject is a daunting one. Even the literature on essential medicines lists and developing medicines policies produced by the WHO is now vast, and requires careful scrutiny and interpretation.

This book aims to provide the student of international health with little previous knowledge of managing pharmaceuticals with an introduction to the international pharmaceutical scene. Who are the key players and what are their roles? What has been achieved so far, and what strategies have been adopted to achieve it? What remains to be done? And what are the prospects for the future?

The book has been written by a group of academics and practitioners from institutions in different countries who collectively have enormous experience of the issues involved. Although each chapter has one or two lead authors in practice all the authors have contributed to all chapters. We think the book has been greatly strengthened as a result. We particularly acknowledge Reinhard Huss's work in developing the medicines cycle, the representation of which appears on the book's cover.

The book will be of interest to all those who wish to learn more about this important area of international health. They will include students and educators from the fields of public health and pharmacy, health professionals, policy makers and programme managers, and indeed all those involved with the day to day problems of improving the state of the world's health by pharmaceutical means.

Stuart Anderson
London, March 2004

About the Authors

Dr Stuart Anderson BSc (Hons), MA, PhD, MRPharmS, MCPP
is a senior lecturer in the Department of Public Health and Policy at the London School of Hygiene and Tropical Medicine, which he joined in 1995. He was previously a lecturer in pharmacy practice at the School of Pharmacy, University of London. He obtained his degree in pharmacy from the University of Manchester. He is a former Teaching Programme Director at the London School, where he had overall responsibility for all masters level education and training.

Early in his career he worked on the development of medicines in the pharmaceutical industry. He has since studied pharmaceutical policy in a wide range of low and middle income countries, and has first hand experience of developments in Brazil, Ghana and Thailand. He is currently president of tropEd, the European Network for Education in International Health, which focuses on improving the management of health services for disadvantaged populations.

Dr Reinhard Huss MD, MPH
is a senior lecturer and consultant in the Department of Tropical Hygiene and Public Health at the University of Heidelberg, which he joined in 1999. He holds qualifications in clinical medicine and public health from German and British universities. He has gained a profound insight into the functioning of health care systems through his diverse experience as clinical doctor, district medical officer, adviser, teacher and researcher in the Central African Republic, Germany, United Kingdom and Zimbabwe.

He focuses on rational medicine management, district health service management (including quality improvement of services), the consequences of health sector reform for district health services, and the importance of human aspects of implementing change. He has participated in the planning, monitoring and evaluation of district and regional health services; the setting up of a regional medicine revolving fund; the development of sexual and reproductive health services; and the facilitation of local SWAp approaches.

Professor Rob Summers, BSc (Pharm), PhD (Bradford)
is currently Head of the School of Pharmacy at the Medical University of Southern Africa (MEDUNSA), which offers the BPharm, MSc(Med) and PhD degrees. He

also manages the Pharmacy Training and Development Project, an independently funded outreach, training and research body with particular focus on peripheral areas. He has published extensively in pharmaceutics, pharmacokinetics, clinical pharmacy and photobiology and has presented papers at many international meetings in Europe, UK and USA.

He has chaired a number of provincial and national committees, including the National Essential Medicines List Committees for Primary Health Care and for Hospitals. He has advised the national Department of Health on the registration, production, procurement and distribution, and overall management of pharmaceuticals and vaccines, including financial management matters. He has directed many national and international short courses on topics ranging from Promoting Rational Medicine Use and Medicine Policy Issues to Rational Medicine Management to Pharmacoeconomics.

Dr Karin Wiedenmayer MSc (Pharm), PharmD
joined the Centre for International Health at the Swiss Tropical Institute in Basel as a senior scientist and lecturer in 1993. She holds a Masters degree in pharmacy from Switzerland and received her doctorate in Clinical Pharmacy from the University of North Carolina, Chapel Hill, USA. She spent several years as a clinical pharmacist in teaching hospitals in the USA. Later she worked with the Swiss Drug Regulatory Authority, as a regulatory affairs and quality assurance manager in private industry, and for an NGO in Calcutta, India.

She has been involved for several years in pharmaceutical sector support, medicine supply and the promotion of rational medicine use. She has an academic appointment with the Faculty of Pharmacy of Muhimbili University College of Health Sciences in Dar es Salaam, Tanzania, and has collaborated with the City Council for the Dar es Salaam Urban Health Project over many years. She has worked as a consultant for the Tanzanian Ministry of Health, the World Health Organization, the World Bank, and the Global Fund. She also coordinates a continuing education programme for Swiss pharmacists.

Acknowledgements

A large number of people have contributed in a great many different ways to the production of this book. We would like to thank all our present and former masters and doctoral students, in London, Heidelberg, Basel and Pretoria, who have contributed in many different ways to its preparation.

In particular Reinhard Huss, Rob Summers and Karin Wiedenmayer would like to thank the many students who have now completed the course on Rational Medicine Management in International Health organised jointly by the Department of Tropical Hygiene and Public Health (ATHOEG) at the University of Heidelberg and the Swiss Tropical Institute (STI) at Basel.

In addition, Reinhard Huss would like to thank Doreen Montag, a PhD student of anthropology, for her help in the preparation of chapters 3, 7, 11 and 12. He would further like to thank all friends, colleagues and partners who have contributed through discussions to the maturation of ideas.

Stuart Anderson would like to thank many friends and colleagues for help and advice. He also wishes to thank past and present students at the London School of Hygiene and Tropical Medicine who have contributed to the development of this venture in different ways. They include Anahi Dreser, Duangtip Hongsamoot, Niyada Kiatying-Angsulee, Ariel King, Naoko Tomita and Helena Walkowiak. Thanks to Martin Anderson for editorial assistance.

Rob Summers would like to thank Monika Zweygarth of the MEDUNSA School of Pharmacy for her immense help in compiling and drafting Chapters 6 and 8, and the many colleagues and friends with whom he has worked in pharmaceuticals management over the years.

Karin Wiedenmayer would like to thank her family, friends and colleagues who supported her during this endeavour, particularly her daughter Anne-Sophie, who wonders what kind of fairy tale this book will become. We suspect this sentiment might be shared by others.

Particular thanks are due to Richard Laing at WHO for his inspiration, support and encouragement for this venture, and for readily agreeing to write a foreword to it. We also thank Trixi Menz at our publishers, Birkhäuser, for her forbearance, gentle prodding and encouragement at all stages of the book's production.

We acknowledge the permissions readily granted by a number of individuals and organizations to reproduce material for which they own the rights. Sources are acknowledged in the text. Stuart Anderson thanks Alan Jones for pre-publication access to one of his articles. Particular thanks are due to the staff of the Essential Drugs and Medicines Policy directorate at the WHO for use of their materials.

Finally, we wish to thank Franziska Matthies, formerly at the University of Heidelberg and currently at the University of East Anglia in Norwich, for her enthusiasm and commitment in bringing the various parties together to further the cause of rational medicine management in international health.

Stuart Anderson
Reinhard Huss
Rob Summers
Karin Wiedenmayer

Glossary of Terms

ACP	African, Caribbean and Pacific countries
ADR	Adverse drug reaction
AHP	Allied health professional
ARV	Antiretroviral
ATC	Anatomical therapeutic chemical (classification system)
ATHOEG	Department of tropical hygiene and public health (University of Heidelberg, Germany)
BNF	British national formulary
CADD	Computer-aided drug design
CAM	Complementary and alternative medicine
CC/HTS	Combinatorial chemistry/high throughput screening
CRO	Clinical research organzation
CSM	Committee on the safety of medicines
DALY	Disability-adjusted life year
DAP	Drug action programme
DGDev	Directorate general for development (of the EC)
DNDi	Drugs for neglected diseases initiative
DPM	Drug policies and management
DRA	Drug regulatory authority
DTCA	Direct-to-consumer advertising
EBM	Evidence-based medicine
EC	European commission
ECHO	European community humanitarian organization
EDI	Electronic data interchange
EDM	Essentail drugs and medicines (policy division)
EDP	Essential drugs programme
EM	Essential medicine
EMEA	European medicines evaluation agency
EML	Essential medicines list
EU	European union
FBO	Faith-based organization
FDA	Food and drug administration
FIP	International pharmaceutical federation

GATB	Global alliance for TB drug development
GATT	General agreement on tariffs and trade
GAVI	Global alliance for vaccines and immunization
GMP	Good manufacturing practice
HAI	Health action international
IAVI	International AIDS vaccine initiative
IBRD	International bank for reconstruction and development
ICH	International conference on harmonisation of technical requirements for the registratrion of pharmaceuticals for human use
ICI	Imperial chemical industries
ICIUM	International conference on improving the use of medicines
IDA	International development association
IDA	International dispensary association
IEC	Information, education and communication
IFPMA	International federation of pharmaceutical manufacturers
IFPW	International federation of pharmaceutical wholesalers
IFRC	International federation of red cross and red crescent societies
IGPA	International generic pharmaceutical alliance
IMF	International monetary fund
INRUD	International network for rational use of drugs
IPRP	Intellectual property rights protection
ISDB	International society of drug bulletins
MMV	Medicines for malaria venture
MRSA	Methicillin-resistant *S. aureus*
MSF	Médecins sans frontières
MSH	Management sciences for health
NACOSA	National AIDS convention of South Africa
NDP	National drug policy
NGO	Non-governmental organization
NMP	National medicines policy
NSAID	Non-steroidal anti-inflammatory drug
OCT	Overseas countries and territories
OECD	Organization for economic cooperation and development
PEM	Prescription event monitoring system
PLoS	Public library of science
PMS	Post-marketing surveillance
PPP	Public-private partnership
PSF	Pharmaciens sans frontières
QA	Quality assurance
QALY	Quality-adjusted life year
QC	Quality control
QSE	Quality, safety and efficacy (standards)
R&D	Research and development
SMO	Site management organization
STD	Sexually transmitted disease

STG	Standard treatment guidelines
STI	Sexually transmitted infection
STI	Swiss tropical institute
SWAp	Sector-wide approach
TBA	Traditional birth attendant
TM	Traditional medicine
TNC	Transnational corporation
TNPC	Trans-national pharmaceutical corporation
TRIPS	Trade-related aspects of intellectual property rights
tropEd	European network for education in international health
UNCTAD	United nations conference on trade and development
UNDCP	United nations drug control programme
UNDP	United nations development programme
UNFPA	United nations population fund
UNHCR	United nations high commissioner for refugees
UNICEF	United nations children's fund
UNIDO	United nations industrial development organization
USP	United States pharmacopoeia
VEN	Vital – essential – nonessential
VRE	Vancomycin-resistant enterococcus
WHO	World health organization
WIPO	World intellectual property organization
WLPS	World list of pharmacy schools
WTO	World trade organization

Chapter 1
Issues in the Management of Pharmaceuticals in International Health

Stuart Anderson and Reinhard Huss

Box 1.1: Learning objectives for chapter 1

By the end of this chapter you should be able to:

· Describe the life cycle of modern medicines.
· List the market share of medicines for each region of the world.
· Describe the academic disciplines that have contributed to knowledge about the use and availability of medicines in international health.
· Explain the benefits of taking a WHO approach to the management of pharmaceuticals in low and middle income countries.
· List key events in the development of the WHO approach to the management of pharmaceuticals.
· List the pharmaceutical components of the level 1 core indicators used in the WHO 2004 "World Medicines Situation" Report.

1.1   Introduction

This chapter aims to lay the foundations for the rest of the book. We have called it "managing pharmaceuticals in international health" only after careful consideration. We begin therefore with an explanation of the terms "pharmaceutical" and "international health" and how we are using them. We then consider medicines in a wider historical and international context, before considering global inequity in access to pharmaceuticals, and the medicines life cycle.

The second half of the chapter is concerned with the study of pharmaceuticals in international health. We examine different approaches to their management, before concluding that the WHO approach is the most appropriate for understanding the issues involved, and in providing practical support and guidance. Finally, we describe the layout of the book and how to use it.

1.2   Pharmaceuticals, medicines and drugs

The products that we are concerned with in this book have been described by many names, and the preferred terms have changed over time. They include "pharmaceutical", "medicine", "medicinal product", "therapeutic drug" and just "drug". Yet these words are not entirely interchangeable, and we need clarify from the outset how they are being used in this book.

Our policy has been to follow the most recent WHO terminology wherever possible, as described in "the World Medicines Situation 2004" and elsewhere. This has effectively completed the transition from preferred use of the term "drug" to use of the term "medicine". "National drug policies" have thus become "national medicines policies", "essential drugs lists" have become "essential medicines lists", and "rational drug use" has become "rational use of medicines".

The reason for this shift is a change in the use and meaning of these words. Over the last ten years or so the term "drug" has increasingly become understood as referring to illicit substances. It has become too inclusive, and too imprecise. Yet substitution of the word "medicine" is by no means universal and is not always appropriate; it continues to appear in many phrases concerned with medicines. These include "drug resistance", "adverse drug reactions", "drug utilization studies" and "drug interactions", amongst many others. And there are agencies and programmes concerned with medicines with the word "drug" in their title, such as the "Drugs for Neglected Diseases Initiative". In all these circumstances we have retained the word "drug". Finally, we have tried to use whatever phrase was correct at the time, so that we use both "Essential Drugs Lists" and "Essential Medicines Lists" depending on the context.

"Pharmaceuticals" is generally considered to have a broader meaning than "medicines". It refers not only to finished medicines but also to active ingredients and vaccines (a group of products of particular significance in international health). For the most part, however, we use these two terms interchangeably. Nevertheless, "pharmaceutical" is the term usually used to describe the industry, and again that is the term we use here. The products of the pharmaceutical industry can be very wide ranging: they include not only bulk ingredients, finished products, vaccines and other biological products, but also over-the-counter medicines, veterinary medicines, diagnostic products and medical devices, amongst others.

"Generics" are pharmaceutical products that are marketed after the expiry of a patent or other exclusivity rights held by the innovator company. They are intended to be interchangeable with the original product. Generics also include pharmaceuticals that have never been patented, and copies of patented medicines in countries that have no such patent. The generics market is itself sub-divided into sub-sectors. "Branded generics" are generic medicines that have been given a brand name of their own. Many manufacturers may market the same generic medicine, each under its own brand name. "Commodity generics", on the other hand, are sold under the generic name. They too may be manufactured and marketed by many different companies.

A distinction also needs to be made between the "research-based pharmaceutical industry" and "reproducer firms". Whilst the former produces a steady stream of new medicines for the market, the latter are involved mainly in the production of generic medicines. The distinction between research-based and generic manufacturer is however frequently blurred: many transnational pharmaceutical corporations themselves have large generic subsidiaries that account for a large share of the world's generic market.

1.3   International health

We have chosen to use the term "international health" in the title of our book in favour of alternatives such as "developing countries". These terms too are not synonymous, and we are aware that there continues to be considerable confusion about the meaning of the term.

International health is a relatively new field of study that systematically compares factors that affect the health of all human populations, with a special focus on poverty-related health problems in low and middle income countries. International health extends beyond the prevention and treatment of diseases, to include the promotion of health, palliative care and rehabilitation. It thus demands a very wide range of knowledge and expertise. This includes knowledge of the major endemic diseases, studies of health systems, health economics, health policy and the management of health services.

Achieving improvements in the health situation in many low and middle income countries is a global challenge. These countries are no longer restricted to the South, the countries of Africa, South America and south-east Asia. They include many other countries often undergoing rapid transition, such as those from the former Soviet Union, Eastern Europe, the Balkan States and countries in the Middle East and central Asia. Trying to reach this goal in such countries presents international health professionals with formidable challenges.

Although the circumstances of each country are usually very different the consequences of lack of access to even basic health care are all too similar. Each year, around ten million children die from preventable infectious diseases such as acute respiratory infections, diarrhoeal diseases, measles and malaria. About one third of the world's population is infected with the organism causing tuberculosis. Around 1.5 million people die each year from tuberculosis; of these deaths 98 per cent occur in developing countries. HIV/AIDS continues to spread rapidly throughout such countries, at the same time as it decreases in industrialised countries.

Other factors contribute significantly to this misery and waste. Both the scope and dimension of disasters and emergencies increase dramatically when they occur in African or Asian countries. They create ideal conditions for the further spread of infectious diseases. And global economic demands have led to structural adjustment programmes that have necessitated the introduction of fundamental reforms to the health sector in many of the poorest countries of the world.

1.4   The emergence of medicines

For most of human history medicines obtained from natural sources were the only ones available. For most of history humans have had to rely on their natural body defences to defend themselves against the barrage of diseases to which they were subject. For centuries the medicines used to assist nature came exclusively from a variety of animal, vegetable and mineral sources. Although some, like opium and digitalis, had very real therapeutic effects, most did little more than offer symptomatic relief at best, or poison the patient (strychnine and arsenic were frequent ingredients of many medicines) at worst.

1.4.1   The development of effective medicines

Although some effective medicines (like acetyl salicylic acid or aspirin) were available before the end of the nineteenth century, the really significant developments, particularly antibiotics and antibacterials, came only in the twentieth century. Some of the research that led to these early discoveries was carried out in university or public laboratories, but the main thrust for medicine research and development came from pharmaceutical companies themselves. There were clearly substantial profits to be made from the marketing of effective medicines for a whole range of society's diseases and illnesses. These companies were largely based in Europe and North America, and their primary target was not surprisingly their own populations.

Right from the beginning the cost of these medicines was a major political issue in their own countries. Many struggled with mechanisms by which the benefits of these new effective medicines could be made available to people at reasonable cost, whilst at the same time ensuring that the pharmaceutical companies had adequate incentives to carry out the research in those countries and to be adequately compensated for taking the risks associated with developing new medicines.

1.4.2   The rise of the pharmaceutical industry

From modest beginnings as "cottage industries" in the mid-1800s many pharmaceutical companies expanded beyond recognition during the second half of the twentieth century. Recent mergers between the largest have led to the creation of a relatively small number of trans-national pharmaceutical corporations (TNPCs). With this has come the rationalisation of research programmes, with particular corporations now focussing their research efforts on a limited range of therapeutic categories. These are private sector organizations whose central objective is the generation of profits for their shareholders. They therefore target the most lucrative markets, which inevitably are those in high and middle income countries. There is a steady stream of medicines constantly under development, the so-called "medicines

Box 1.2: Sales of medicines in top ten therapeutic classes 2001

Class	Total sales (US $ billion)	per cent share of global sales
Anti-ulcer	19.5	6
Cholesterol and triglyceride reducers	18.9	5
Antidepressants	15.9	5
Non-steroidal anti-inflammatory drugs	10.9	5
Antihypertensive drugs (Ca antagonists)	9.9	3
Antipsychotics	7.7	2
Oral antidiabetics	7.6	2
ACE inhibitors (plain)	7.5	2
Antibiotics (cephalosporins and combinations)	6.7	2
Systematic antihistamines	6.7	2
All Ten	111.3	32

Source: The World Medicines Situation, 2004, Geneva: WHO, Table 1.3.

pipeline". These may well offer advantages over existing therapies, but they will also be more expensive.

Global sales of medicines in the top ten therapeutic classes are illustrated in Box 1.2. These therapeutic classes represent the diseases of affluence: diabetes, high blood pressure, stress, ulcers and depression. They account for over 30 per cent of global pharmaceutical sales, and sales of the ten best selling medicines account for US $ 40.2 billion, or 13 per cent of global market share. In value terms, ten countries account for 85 per cent of all pharmaceutical production, and ten companies account for about half of all sales.

At the moment these new medicines are still of limited interest to those in developing countries. However, an epidemiological transition is taking place in many low and middle income countries so that increasingly we observe diseases of poverty and affluence side by side in these countries. Whatever the situation these countries are largely dependent on the global pharmaceutical companies to come up with the medicines required to fight AIDS, malaria, and tuberculosis. The latest medicine to reduce blood cholesterol levels, or to help fight obesity, has no relevance to those struggling to survive. Their needs are far more basic, but the way pharmaceutical research is structured and funded throws them into stark contrast. On the one hand they need expensive new medicines (such as the antiretrovirals) to fight killer diseases like AIDS and drug-resistant tuberculosis and malaria; on the other hand they need old but effective medicines in sufficient quantities to make a difference to their lives and prospects for survival.

1.5   Global inequity and the medicines life cycle

Modern medicines have a natural life cycle, which is illustrated in Box 1.3. The cycle usually begins with the identification of need for a particular pharmaceutical; market analysis is undertaken before research and development begins; once a likely candidate is found it is patented, and market authorisation and licensing obtained; production and marketing then take place, resulting in the medicine's selection; it is then usually prescribed and dispensed, or distributed and supplied before eventually being taken by a patient. Finally, it may reach a state of obsolescence to be replaced by a new product.

1.5.1   Medicines and research and development

Global inequity in access to medicines occurs at all stages in the life cycle. Research and development of medicines for poor people is largely neglected by large pharmaceutical companies, because these products are unlikely to generate substantial profits. As one industry spokesman has said "The pharmaceutical business, like every other business, is about making money, and it is towards the goal of making the most possible money that industry executives normally devote their energies day in and day out". During the twenty-five years between 1975 and 1999, a total of 1,393 new chemical entities were marketed somewhere in the world. However, only sixteen of these were for tropical diseases and tuberculosis. A newly discovered molecule is unlikely to be developed for human use if a pharmaceutical company anticipates that sales will not be sufficient to make a worthwhile profit.

When pharmacological research is successful, a new medicine may be patented, licensed and regulated in one or several countries. However it is important to differentiate between patenting, marketing authorization, licensing and registration. The terms "marketing authorization" and "license" are often used interchangeably. "Marketing authorization" means that a national medicine regulatory authority has granted the right to sell or distribute a product in a country after assessment of the product. Although the term "license" is often used in the same way, it can also mean that the holder of a patent grants the right to exploit a patent to a third party.

1.5.2   Medicines and patents

A "patent" is a state-guaranteed temporary monopoly for the exploitation of an invention granted to a company or a person who explains the invention and claims the monopoly. International agreements on pharmaceuticals were consolidated following institution of the World Trade Organization (WTO) in 1994. An annexed agreement on Trade-Related Aspects of Intellectual Property Rights (TRIPS) defined the patent life of a pharmaceutical product as twenty years from the date of registration. International guarantees of monopolies on medicines have been hotly contested ever since, and are seen by many as a global reinforcement of inequity.

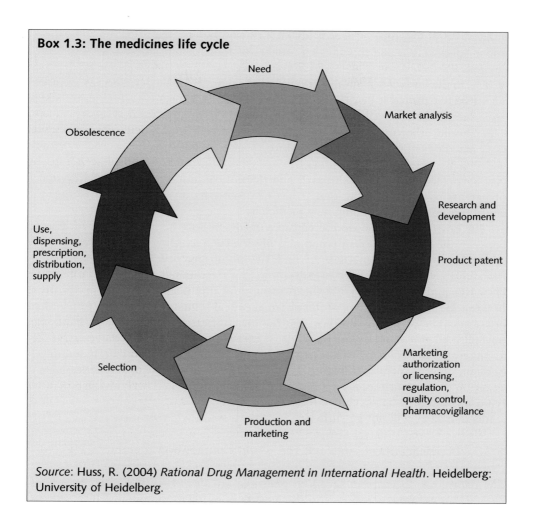

Box 1.3: The medicines life cycle

Need

Market analysis

Obsolescence

Research and development

Use, dispensing, prescription, distribution, supply

Product patent

Selection

Marketing authorization or licensing, regulation, quality control, pharmacovigilance

Production and marketing

Source: Huss, R. (2004) *Rational Drug Management in International Health*. Heidelberg: University of Heidelberg.

International discussions about the TRIPS agreement in recent years have generated considerable interest by both the public and organizations concerned with the supply of essential medicines. The 1999 WHO document "Globalization and access to medicines" informed both governments and international civil society that these new patent rules could be overruled under certain circumstances: these included public health emergencies such as the HIV/AIDS pandemic.

Many commentators question the wisdom of granting patents on pharmaceutical molecules at all. The whole concept of patents has been a matter of dispute in some industrialized countries for many years. The Netherlands introduced and then abolished patents whilst Switzerland refused to introduce them in the nineteenth century following widespread recognition of the problems associated with state-guaranteed monopolies. There is an important difference between an "invention" and a "discovery", and this raises a number of fundamental questions:

· Can the knowledge about a pharmaceutical molecule be described as an invention, or should it be defined as a discovery?
· Should the molecule be patentable, or only the process for making it?
· Should other options for funding be considered in order to encourage pharmacological research based on global health needs?

Not surprisingly, the research-based pharmaceutical industry defends the existence of patents robustly, arguing that it has to protect its investments and recoup the enormous costs of developing new medicines.

The costs involved in developing new medicines are a further issue of dispute between the industry and international observers. According to the industry these costs amount to between US $ 0.5 and 1.0 billion per new product. Independent researchers and civil society organizations have estimated the figure at between US $ 100 and 200 million. At the present time this dispute cannot be resolved, because the financial figures of the pharmaceutical industry are neither transparent nor accessible for public scrutiny. But the huge difference in the cost estimates can probably be explained by the methods of cost accounting used. Industry estimates are likely to include costs associated with marketing as well as research and development, and the inclusion of so-called "opportunity costs" of capital, reflecting the situation if the money spent on research and development had been invested in other types of business.

1.5.3 Pharmaceutical sales and the world's population

The enormous disparity in access to pharmaceuticals is best illustrated by a comparison between the percentage of the world's population represented by a particular region, and the proportion of pharmaceutical sales accounted for by that region. The figures are presented in Box 1.4. North America, with only 6 per cent of the world's population, accounts for over fifty per cent of total world sales of pharmaceuticals. Asia, Oceania and Africa, which together account for 71 per cent of the world's population, are the recipients of only eight per cent of total world sales of pharmaceuticals.

Even allowing for the very high price of pharmaceuticals in North America, together with a reliance on expensive medicines to treat the diseases of an affluent life-style, like those for obesity, depression and high blood pressure, these figures demonstrate a very high level of both inequality (lack of equivalence) and inequity (lack of fairness), if we believe that vital commodities such as pharmaceuticals should be provided according to the health needs of the population rather than ability to pay.

Available medicines are often too expensive and therefore inaccessible for the global poor. This problem of availability and distribution is particularly obvious

Box 1.4: Global population and pharmaceutical market share

Region	Population (%)	Pharmaceutical sales (%)	Value of sales (estimates)* (US $ billion)
North America	6	51	204
Europe	12	25	100
Japan	2	12	48
Asia, Oceania, Africa	71	8	32
Latin America	9	4	16

Notes: Sales cover direct and indirect pharmaceutical channel purchases in US dollars from pharmaceutical wholesalers and manufacturers. The figures represent 52 weeks of sales data, and include prescription and certain OTC data and represent manufacturer prices.
Website: http://www.ims-global.com/insight/news_story/0302/news_story_030228.htm
*Based on the estimate of US $ 400 billion world wide sales
Sources: Deutsche Stiftung Weltbevölkerung, DSW Website: http://www.dsw-online.de/ and IMS World Review 2003.

during natural and man-made disasters. However, the enormous costs of medicines are also a major concern in most industrialised countries, where health care spending generally grows at a faster rate than the overall economy. Pharmaceuticals are one of the most important contributors to this cost explosion and take up an increasing share of health care expenditure. But the difference between industrialised and developing countries in relation to medicines is stark: for industrialised countries it involves difficult choices; for developing countries it involves countless unnecessary deaths.

Addressing the gross inequities in access to pharmaceuticals requires positive action. Many current developments mitigate against it. Many of these developments aim to create a uniform global market for pharmaceuticals. The process of standardizing licensing and regulatory requirements for medicines was begun in 1991 with the first International Conference on Harmonisation of Technical Requirements for the Registration of Pharmaceuticals for Human Use (ICH) in Brussels. This process is supported by the regulatory authorities of the USA, Japan and the European Union, as well as the pharmaceutical trade associations. It aims to rationalise medicine testing and to speed up the marketing of new medicines for the industrialized regions of the world. It is argued that this process may further undermine pharmaceutical development in low income countries.

1.5.4 The rational use of medicines

The use of medicines by patients comes at the end of a highly complex chain of events involving immense financial investment, starting with research and development, continuing to manufacturing and quality control, regulatory approval and marketing, through to selection and procurement, distribution and storage, to the final therapeutic consultation and the actual administration of a medicine. However, the whole supply chain will be of little point if it is not followed by rational medicine use.

Irrational use of medicines is a global problem. In countries with limited resources it significantly adds to economic losses, and contributes to increased morbidity and mortality. The result of inefficiencies, waste and irrational medicines use is that far more is spent on pharmaceuticals than is necessary. So much can be achieved by making more effective use of existing resources. The performance and quality of health services everywhere critically depends on the availability and rational use of medicines.

For 25 years WHO, along with other international organizations, has advocated the concept of "essential drugs", now called "essential medicines", and has promoted and supported national essential medicines programmes. Initially much energy and resources went into improving medicine supply systems in order to increase the *availability* of medicines. At the Nairobi Conference of Experts in 1985, the need for more information on the medicine situation at the global and national levels was recognised, and in 1988 WHO published a survey of the "World Drug Situation". This survey rated the state of development of ten aspects of pharmaceutical management in 104 low and middle income countries.

Monitoring medicines policies is a complex task. Since 1988 a great deal of effort has gone into the development of indicators and the collection of country-wide data. A number of monitoring tools are now available, mainly indicator-based. The 1994 WHO manual *Indicators for monitoring national drug policies* listed around 120 indicators covering structure, process and outcome elements of various policy components. The indicators were subsequently updated in 1999 (WHO/EDM/PAR/99.3). These have been extensively used since, and it is these that we describe more fully in chapters 6 and 11.

The 2004 WHO package *Core indicators for monitoring and assessing country pharmaceutical situations* takes the process a stage further. The package takes a hierarchical approach to monitoring based on three groups of core indicators, the data for which can be easily collected using simple survey techniques. Level 1 core indicators assess existing structures and processes in national pharmaceutical systems. Ten aspects of the systems are assessed; these are listed in Box 1.5. For each a varying number of questions are asked, usually of a knowledgeable informant within a ministry of health. Examples only of these questions are given in the Box. Level 1 indicators provide a useful benchmark for monitoring progress at the national level in the management of pharmaceuticals in international health. We return to indicators in later chapters.

Before 1985 little attention was given to how medicines were actually used. But the Nairobi Conference recognised the importance of the *rational use of medicines*: the availability of essential medicines was futile if they were not used rationally. The importance of ensuring rational medicine use has since become a priority in international health, and much has already been done to improve the ways in which

Box 1.5: WHO country level indicators 2003

Situation in country	Examples of indicator questions
1. National medicines policy	Is there a national medicines policy (NMP) document? Is the NMP integrated into a published national health plan? Is there a national policy on traditional medicine?
2. Legislation/regulation	Is there a medicines law? If yes, when was it last updated? What is the system and operation of medicines registration? Are there written national guidelines for inspection?
3. Quality control	How many medicine samples have been tested? Where have these samples been tested? How many failed identity or assay?
4. Essential medicines list (EML)	Do essential medicines lists exist? Where are they being used? Are local herbal medicines included on the national EML?
5. Medicines supply system	Who is responsible for public sector drug procurement? Is government procurement limited to medicines on the EML? What type of tendering process is used?
6. Medicines financing	What is total government budget for medicines in US $? Which medicines are free at primary public health facilities? What proportion of the population is covered by insurance?
7. Access to essential medicines (EMs)	What proportion of the population has regular access to EMs? What proportion is within a 1 hour walk of facilities that have EMs available? What proportion can afford EMs?
8. Production	What is the medicines production capability in the country? What is the value of annual sales in US $ for each category? What is the total volume and value of the medicines market?
9. Rational use of medicines	Are there standard treatment guidelines available? Is there a national strategy to contain antimicrobial resistance? Are antibiotics and injections sold over the counter?
10. Intellectual property rights protection (IPRP) and marketing authorization	Is patent protection provided for pharmaceutical products? What protection is provided for traditional knowledge? Is your country a WTO member?

Source: WHO (2003) Operational Package for Monitoring and Assessing Country Pharmaceutical Situations. Geneva: World Health Organization.

medicines are used by prescribers, dispensers and patients. The First International Conference on Improving the Use of Medicines (ICIUM) was held in Chiang Mai, Thailand in 1997. This conference brought together participants from many countries and a wealth of important initiatives and ideas, and provided the impetus for a wide range of developments aimed at improving the rational use of medicines. A Second ICIUM conference was held in 2004.

1.6   The study of pharmaceuticals in international health

Pharmaceuticals have been a major focus of intense investigation and research for many different reasons. The global pharmaceutical industry is today a powerful force in the world's economy. It employs large numbers of people, shares in it are owned by millions of others, it contributes substantial sums to the exports of industrial countries, and it generates large sums in taxation. In short, it is economically important. In this, it is not unlike other industries such as aerospace, computing or motor manufacturing.

But pharmaceuticals are not simply commodities like any other. People can manage very well without cars, computers and aeroplanes, but lack of access to affordable medicines may well mean a death sentence. It is their central importance in health care and public health that makes them of interest to a far wider group of researchers and commentators than others. They have a significance that goes far beyond their economic value. They go to the very heart of debates about equity, poverty and health across the world. These issues are of concern to many people from many different backgrounds.

As a result a number of different perspectives have been taken. For example, health professionals from a wide variety of backgrounds have an involvement in the management of pharmaceuticals in international health. These include not only doctors, nurses and pharmacists, but also health service managers, supplies officers and finance officers. Their perspective is usually formed from experience in trying to provide the best possible health care to as many people as possible.

The role of pharmaceuticals in health care is so fundamental that not surprisingly it has been a subject of interest to a very wide range of academics from many different disciplinary backgrounds. In addition to public health, medicine and pharmacy, key ones are health economics, medical sociology, epidemiology, medical anthropology and health psychology. All these disciplines have contributed to our knowledge about the need for and use of medicines.

The literature in this field is now vast, and for those coming new to the area the prospect of making sense of it can be daunting. By far the largest literature relates to the use of pharmaceuticals in industrialised countries. There is an enormous literature that relates to the pharmaceutical industry: there is a vast literature concerned with the taking of medicines, with compliance and concordance; and there is a large literature around the policy issues relating to access to pharmaceuticals in developing countries. WHO itself now has a substantial literature about access to and the use of medicines, and there is a growing literature around cultural differences in attitudes to and uses of medicines. However, a number of approaches to the management of pharmaceuticals in international health have emerged, and we will now consider them further.

1.7 Approaches to managing pharmaceuticals

With so much published and with so much literature there are obviously many ways in which the subject can be approached. Indeed, there are many books that cover particular aspects of the subject from a particular discipline or perspective. In an introductory book like this one our aim is to help the reader understand the different perspectives, to make intelligent judgements about much that is published about the issues and the industry, and to have a clear framework in which these can be taken forward.

1.7.1 General approaches

The management of pharmaceuticals in international health has been considered and studied in some depth at a variety of levels in the health care delivery chain. These range from the local community and district level, up to the national, regional and international levels. All have their place, and all are important if whole populations are to have access to the full range of effective medicines, and their rational use is to be achieved.

1.7.2 The WHO approach

We have found it most helpful and productive for students of pharmaceutical policy in international health to take a perspective grounded in the experience of WHO. There are a number of important reasons for taking this perspective:

· WHO has been grappling with the issues for over thirty years, with the essential medicines programme first established in 1977. It thus has extensive experience of them.
· The WHO approach is founded on providing sound practical advice to countries about the needs and priorities concerning access to and use of medicines.
· This guidance is freely available to anyone, and can be downloaded from the WHO website. This provides a valuable resource to both practitioners and students.
· The essential medicines programme deals with the many and complex issues which arise in a systematic way, which enables subsequent developments to be absorbed and accommodated.

1.8   Layout of the book

We have attempted in this book to give a brief introduction to all the key issues, approaches and tools of which students interested in this field will need to be aware. The book is divided into four main sections.

1.8.1   Access, need, and supply

We begin in chapter 2 by describing the central problem in managing pharmaceuticals in international health, that of access to pharmaceuticals and the availability of appropriate medicines. This chapter also addresses the issue of neglected diseases, and considers strategies for developing new treatments and ensuring access to existing ones.

In chapter 3 we consider how we might quantify the problem, by providing an account of how we assess the medicine needs of patients and populations. This is by no means straightforward, since need is as much influenced by the cultural aspects of the environment in which medicines are taken as by the therapeutic need as defined by clinicians.

One of the great influences on patients regarding the use of medicines is the people closest to them. After their immediate family and communities these are the health professionals. In chapter 4 we consider the classification of health professionals and consider the roles of key groups in the prescribing, dispensing and administration of medicines taking. We include here the place of traditional medicines and the role of traditional healers.

In recent years the role of the global pharmaceutical industry has tended to dominate discussion about access and availability of medicines in international health. In chapter 5 we describe the origins, structure and activities of the industry, before considering the nature of pharmaceutical markets and the place of the pharmaceutical industry in international health.

1.8.2   The actors in international health

The following three chapters consider the roles of the major players involved in managing pharmaceutical use in international health. In chapter 6 we explore the crucial role played by individual governments. Here we discuss the development of national medicine policies, the medicine supply process, and the importance of health systems that support medicine supply. We move on to the role of governments in pharmaceutical legislation, regulation and enforcement.

There are many key players on the international stage that have a major influence on pharmaceutical use in international health. These include the European Union, national assistance agencies, and both national and international NGOs. These are considered in chapter 7. We consider the contribution that these agencies can make to equitable pharmaceutical supply and distribution.

In chapter 8 we consider the role of other international organizations that have a major role in the management of pharmaceuticals in international health. Prominent amongst these is the WHO, but we also review international organizations representing the pharmaceutical industry, amongst others. In this chapter we also consider the concept of essential medicines, discuss pharmaceutical procurement systems, and explain international agreements and intellectual property rights affecting pharmaceuticals.

1.8.3   Medicines and their use

The next four chapters are concerned with the medicines themselves. International initiatives to improve access to pharmaceuticals is wasted if those medicines are not then used rationally. Chapter 9 describes the importance of rational medicine use. It demonstrates the consequences of the non-rational use of medicines, and considers medicine use behaviour. It explains how medicine use problems can be investigated, and describes interventions that can be used to improve rational use. Finally it reviews the effectiveness of those interventions.

In chapter 10 we consider the issues around the quality assurance of pharmaceuticals, and explore the problems of substandard and counterfeit medicines. We discuss measures that can be put in place to reduce the incidence of poor quality medicines. We also consider in this chapter the fact that medicine use, whether rational or non-rational, often has a number of unintended consequences. The first of these is antimicrobial resistance, resulting from inappropriate use of medicines; and the second is adverse medicine reactions.

Crucial to the achievement of the rational use of medicines is the availability of good information about the actions and uses of medicines. In Chapter 11 we describe the users of medicines information, and the principal sources of it. We examine the quality of this information and describe ways of assessing it. Finally in this chapter we describe the systems that are in use for the classification of medicines.

All these many initiatives may or may not be having an impact on access to medicines, on their rational use, and on the improvement in health status of populations. We need to know whether we are making progress or not. If things are to be improved we need to know what works and what doesn't. This involves research on medicine use, and this is the subject of chapter 12. Here we describe medicine utilization studies and the various disciplines of pharmacoeconomics, pharmacovigilance and pharmacoepidemiology.

1.8.4   Trends, prospects and implications

In our final chapters we review progress overall. Are things getting better or worse? What are the challenges that lie ahead, and what strategies will be needed to achieve the level of access and availability of medicines in international health that are so desperately required? In chapter 13 we consider the prospects of pharmacogenetics and pharmacogenomics, the contribution of plant medicines, the role of the internet, and changes in the roles of health professionals.

Finally, in chapter 14, we examine the changing role of public-private partnerships, the roles of the Global Forum for Health Research and the Global Fund to fight AIDS, tuberculosis and malaria, and the global procurement of medicines. We conclude with a brief review of the implications of recent policy initiatives in the field of pharmaceuticals in international health for the industry, for governments, and for the people of the world.

1.9   Conclusion

In this chapter we have set the scene for the contents of this book. We think we have laid it out logically, and we suggest that readers start at the beginning and work through it. However, the individual chapters contain much valuable detail about the management of pharmaceuticals in international health, and we think that readers will also find it to be a valuable reference source, which they will want to return to again and again as they become more involved with this complex but rewarding field.

We have headed each chapter with a number of learning objectives, which we hope will be helpful in structuring learning and focusing on key concepts and initiatives. We have included at the end of each chapter up to a dozen references for further reading. We have limited these to material that is readily accessible in the English language. We have in addition included a list of useful websites at the end of the book.

Further reading

Abraham, J. and Lewis, G. (2000) *Regulating Medicines in Europe*. London: Routledge.

Bruden, P., Rainhorn, J.D. and Reich, M. (1994) *Indicators for Monitoring National Drug Policies*. Geneva: World Health Organization.

Chetley, A. (1995) *Problem Drugs. Health Action International*. London: Zed Books.

OECD Health Data 2003 (2003) Second edition. Paris: Organization for Economic Cooperation and Development.

Pecoul, B., Chirac, P., Trouiller, P. and Pinel, J. (1999) "Access to essential medicines in poor countries: a lost battle?" *Journal of the American Medical Association* 281: 361–367.

Quick, J.D., Rankin, J.R., Laing, R.O., O'Connor, R.W., Hogerzeil, H.V., Dukes, M.N.G. and Garnett, A. (eds.) (1997) *Managing Drug Supply: The Selection, Procurement, Distribution and Use of Pharmaceuticals*. Second edition. West Hartford, CT, USA: Kumarian Press.

Schwartz, H. (1991) "Improving the industry's image". *Scrip* 1620/21: 22–23.

Vogel, D. (1998) "The globalization of pharmaceutical regulation". *Governance* 11: 1–22.

WHO (1988) *The World Drug Situation*. Geneva: World Health Organization.

WHO (1999) *Globalization and Access to Medicines*. Health Economics and Medicines DAP Series No. 7. Geneva: WHO/DAP/98.9. Geneva: World Health Organization.

WHO (2003) *Operational Package for Monitoring and Assessing Country Pharmaceutical Situations*. Geneva: World Health Organization.

WHO (2004) *The World Medicines Situation*. Geneva: World Health Organization.

Chapter 2
Access and Availability of Pharmaceuticals in International Health

Karin Wiedenmayer

Box 2.1: Learning objectives for chapter 2

By the end of this chapter you should be able to:
- Describe the global drug gap.
- List reasons for inequalities in access to essential medicines.
- Define accessibility, availability, acceptability and affordability in relation to medicines.
- Describe four strategies for improving access to medicines.
- Define neglected diseases and most neglected diseases.
- List factors leading to diseases being neglected.
- Describe three strategies for developing treatments for neglected diseases.

2.1   Introduction

Millions of people worldwide still do not have access to essential medicines that are affordable and of good quality. Access to medicines means access to treatment. Improving access to quality treatment is currently the most important strategy to reduce death and disability from many diseases. More generally, ensuring access to effective treatment is a high priority issue for international public health. Over one third of the world's population lacks access to medicines for many reasons, all of which must be addressed in a comprehensive manner. The most important is poverty, which means that neither the poor nor their governments can afford to purchase essential medicines or ensure their rational use in well-run health systems. Affordability is one of the central issues in debates about medicine use in international health.

There are however other major factors which deny the populations of developing countries access to effective medicines for the treatment of the diseases to which they are subject. Poor infrastructure and unreliable medicine supply systems, and waste and inefficiencies in managing logistics, add to low availability of medicines. This will be discussed in a later chapter. Although basic medicines research is carried out by both the public and private sectors, global pharmaceutical companies

devote greater effort and funding to the development of products which are likely
to bring in profits for their shareholders. This objective means that they focus mainly on the diseases of developed countries, with the result that diseases prevalent in
developing countries are largely neglected.

Many of the issues surrounding the accessibility of medicines in developing
countries can only be addressed with concerted international action. In this chapter
we consider the issues of access and availability of pharmaceuticals in international
health, and describe some of the initiatives that have been taken to address them.

2.2   Access to essential medicines

Medicines are a significant part of health care: they can save lives and improve
health. They promote trust, involvement and participation in health services. Medicines are key factors for an effectively functioning health care system. In fact, medicines are one of the most cost-effective elements of modern health care. However,
not all medicines represent value for money, and often medicines are marketed with
little concern for the real needs and priorities of the people, particularly in developing countries.

Lack of access to medicines has been described as the *Global Drug Gap*. This
description states that global inequities in access to medicines exist between rich and
poor countries because of market and government failures, as well as huge income
differences. Multiple policies are required to address this global drug gap. We begin
by considering the nature of the inequalities in access to pharmaceuticals.

2.2.1   Inequities in access

Illness is a major reason why poor populations remain trapped in poverty. Yet cost-
effective tools for the treatment of disease exist. Medicines are one important tool
to improve health and secure development gains. However they have to be available
and affordable by all people. There is a high degree of inequity in medicine accessibility. Up to 30 per cent of the world's population lack regular access to medicines.

The TRIPS agreement by the WTO provides twenty years of patent protection
for the originators of new medicines, with the goal of promoting innovation by providing incentives to invest in research and development. From a public health point
of view, however, one consequence of this is that the introduction of affordable
generics is delayed. This is a major hindrance to medicine access, and has been of
particular concern in relation to the availability of those medicines that provide
treatment for the pandemic of HIV/AIDS.

Many medicines on the world market are of poor quality, and are also unaffordable to the vast majority of a country's population. Medicine use is often irrational, all over the world. Global spending on health research by both the public and
private sectors amounted to about US $ 70 billion in 1998. Global medicine spend-
ing as a percentage of total public and private health expenditure varies between less

---

Box 2.2: Need for access to antiretroviral drugs

It is estimated that more than 40 million people worldwide are infected with HIV, and more than 90 per cent of infected people live in the developing world in 2003.

WHO estimates that nearly six million people living with HIV/AIDS need access to care and support including antiretrovirals (ARVs). However, fewer than 5 per cent of those who require treatment in developing countries can access these medicines.

Left untreated, HIV infection results in a period of 3 to 10 years or more during which the infected person remains relatively healthy. Once symptomatic disease or AIDS develops, without access to antiretroviral treatment, death results within an average of two years.

In high-income countries, an estimated 1.5 million people live with HIV, many of them productively, due to pervasive antiretroviral therapy. In the USA, the introduction of triple combination antiretroviral therapy in 1996 led to a decline of 70 per cent in deaths attributable to HIV/AIDS.

Source: WHO/EDM website. (2003) http://www.who.int/medicines/

---

than 10 per cent in industrialized countries to up to 60 per cent in developing countries. In developing countries up to 90 per cent of medicine expenditure is paid for out of the pockets of the patients themselves.

There is a huge gap between the significant health impact of medicines and the reality for millions of people. For them, medicines remain unavailable, unaffordable and unsafe, and even when available they are often used irrationally. A range of initiatives is required to ensure rational medicine use. We address these in a later chapter.

Whilst it is important that we quantify these problems, numbers alone cannot reveal the enormous suffering and real tragedies that affect the daily lives of millions of people. Nowhere is this better illustrated than with the antiretroviral medicines used to treat HIV infection, as indicated in Box 2.2.

2.2.2 Defining the problem

Access to medicines is a multidimensional concept. An analytical perspective reveals not only the complexity of the issue but also the multi-faceted definitions and approaches used to investigate it, and the diversity of solutions recommended to overcome lack of access. What is needed is a broad framework that integrates perspectives from various disciplines.

Access to essential medicines correlates closely with other measures of health system performance such as access to health care, health outcomes, health systems responsiveness, fairness in health financing and income level. Causes of medicine deficiencies can be summarized as poverty and resource constraints, unaffordable medicine prices, limited human resources, restrictions imposed by TRIPS and patent laws, limitations of infrastructure, weak logistics systems, poor regulatory capacity and lack of leadership.

An important outcome indicator for access is whether people get the care they need. Although treatment is the final goal of seeking care, it is only one aspect. Care also involves other elements ranging from communication to diagnosis and medicine quality. Health policy and broader social, economic and political forces affect both access and care. The interaction between them can be studied by examining four distinct but related dimensions of access. These are:

- *accessibility* (ability to reach facilities and obtain services in a real and figurative sense),
- *availability* (actual organization and provision of services),
- *acceptability* (trust in technology and competence of the provider), and
- *affordability* (direct and indirect costs of using the services).

Thus access alone is not enough. We also need to ask whether the care people get is *effective*: are health care services and interventions actually being delivered according to defined standards and respecting patient expectations?

We can examine these dimensions of access in terms of the relationships between them, and from the perspectives of both medicine provision and the patient. The provision of medicines takes place within a much broader system of health care service delivery. Medicines and service delivery therefore need to be considered together. Service delivery includes not only information, advice, instructions and precautions, but also interpersonal attitudes such as politeness and friendliness. Indicators have been developed to measure and monitor these dimensions of medicine access.

To investigate whether people have access to the quality care they need we have to consider the interaction between two sets of characteristics: 1) individual and household factors, and 2) health care system factors. These are outlined in Box 2.3.

Individual and household factors: A review of available data and experience indicates that management of diseases is often highly situational, dynamic and complex. Personal, cultural, social, economic and situational factors combine to influence access to effective care. Of particular importance is the interplay of acceptability and availability.

Health system factors: Accessibility and affordability are clearly important obstacles to treatment. In addition, in many developing countries, weak health care infrastructures and inadequately trained staff often lead to incorrect diagnosis and treatment. Inadequate logistics and difficulties of physical access further compound the problem.

Another way of looking at access is to examine determinants and indicators. These are summarised in Box 2.4.

Quality and safety of medicines, as well as service delivery, are preconditions for effective medicine therapy, and are therefore overarching requirements for medicine access. However, the determinants of access to medicines are many and reflect the complexity of a health system. By looking at these determinants we can begin to

Box 2.3: A framework for the study of access to effective treatment and care

Individual and household factors	Barriers to access	Health system factors
Skills in raising social and economic resources Literacy, education, gender Household circumstances	Accessibility	Spatial distribution Oral and written information
Alternative options (home treatment, traditional healers, drug retailers)	Availability	Opening hours and waiting times Equipment and medicines
Personal, social and economic resources Time (e.g. competing demands and activities)	Affordability	Direct cost (prices, cost sharing) Indirect cost (loss of working days, income) Hidden cost (bribes)
Cultural knowledge about signs, causes and severity of illness Trust in technology and competence of provider	Acceptability	Staff motivation and friendliness Supervision and quality control Interactions with other providers
	Quality of Care	
Expectations by patient	Quality of clinical care (diagnosis and treatment)	Training, motivation Supervision and support Drug resistance and quality of drugs
Assertiveness, knowledge and communication skills	Inter-personal care	Staff motivation to listen and communicate Communication skills

design interventions to improve access. We return to the issues surrounding the quality and safety of medicines in a later chapter.

2.2.3 Strategies to improve access to medicines

Enhancing access to medicines needs concerted action, and can only be tackled with the commitment of all the actors involved. WHO has proposed a framework of complementary approaches for collective action. The four components of this framework are:

Box 2.4: Determinants and indicators of access

Dimension	Definition	Indicators	Determinants
Geographic accessibility	Location of the medicine/service and location of the patient.	Travel time to health facility* Operating hours of health facility	Health facility and patient location Transport Human resources
Availability	Type and quantity of the medicine/ service needed and the type and quantity of the medicine/service provided	Percentage of key medicines available Type and quantity of medicines	Medicine demand and medicine supply Medicine supply management Staff capacity and performance
Affordability	Price of the medicine/ service provided and the patient's ability to pay for the medicine/service	Number of working days for average citizen to pay for full treatment	Income and prices
Acceptability	Attitude and expectations about the medicine/ service and the actual characteristics of the medicine/service	Patient satisfaction	Cultural and psycho-social attitudes and beliefs

*Access has been defined as a 2 hour walk or 10 km distance to a health facility that stocks 25 essential medicines.
Source: INRUD/WHO (2003) http://www.msh.org.inrud/

· rational selection,
· affordable prices,
· sustainable financing, and
· reliable health and supply systems.

Other strategies that are needed to support these components include national drug policies, public-private partnerships and reorientation of the research agenda. The main elements of each of these strategies are summarised in Box 2.5. We explore them in more detail in later chapters.

A new and promising access model for medicine supply is the franchise model, with franchised medicine outlets owned by trained community health workers and inspected by health authorities. This provides a good example of how a public-private partnership can work effectively in pharmaceutical distribution.

Box 2.5: Strategies for improving access to essential medicines

Strategy	Goal	Elements
Rational selection	To define what is most needed	Evidence-based and cost-effective Treatment guidelines Essential medicine lists, formularies
Affordable prices	To adapt to purchasing power of patients and allow for return on R&D investments	Competition Price information Generics Bulk purchasing Negotiation on price controls Sound supply management Reductions of taxes and duties Differential pricing TRIPS safeguards
Sustainable financing	To increase sustainable funding	Government funding Health insurance Community financing NGO support Employer health financing Targeted external funding
Reliable health and supply systems	To ensure quality and availability	Health facility coverage Efficient medicine supply management Public-private-NGO mix Efficient medicine regulation Quality assurance GMP Combating counterfeit and sub-standard medicines

Source: WHO/EDM (2003) http://www.who.int/medicines/

2.2.4   Campaigns and initiatives

Access to medicines in developing countries is a highly political issue, and as such it is subject to intense lobbying by all the principal stakeholders. Leading NGOs such as *Oxfam* and *Médecins sans Frontières* have run vigorous campaigns advocating that all people should have access to essential medicines, especially for the treatment of the main killer diseases: malaria, HIV/AIDS and TB. Many organizations and institutions working in poor countries have put "treatment access" at the top of their agenda. But "access for all" remains a challenge that is difficult to achieve.

Access initiatives run by the NGOs, other advocacy groups and UN organizations have had a significant impact on public opinion, media reporting, international organizations and the pharmaceutical industry. These campaigns have also raised important questions about social responsibility and ethical dimensions. Because most medicine research and development is carried out by a for-profit pharmaceutical industry, a central issue is whether, or at least to what extent, private shareholders should benefit financially from the ill-health of those who need medicines, at the expense of those who cannot afford them and who may well die in their absence.

The fundamental question of "corporate profit or public health?" has opened up an important and critical debate on corporate social responsibility, and on the interrelation of all the stakeholders involved in health systems in a globalized yet pluralistic world.

2.3   Neglected diseases

People in developing countries represent about 80 per cent of the world's population, but only about 20 per cent of worldwide medicine sales. Medicine research and development for developing countries and for diseases of the poor remains very limited, despite an ever-increasing need for safe, effective and affordable medicines for the treatment of these neglected diseases. Diseases of the developing world remain largely un-addressed, because there are few financial incentives for large pharmaceutical companies to address them.

2.3.1   Identifying and defining the problem

Most of the world's population has not benefited from the medical revolution of the last decades. An imbalance between their needs, the medical potential and the actual availability of medicines has led to immense suffering and poor health outcomes. The rest of the world has largely ignored the negative economic and developmental impact of this neglect.

Sleeping sickness afflicts up to 500,000 people per year and threatens another 60 million in Africa alone. Until recently patients had to undergo painful treatment with an arsenic-based medicine because more effective treatment was unavailable. A quarter of the population of Latin America is threatened by Chagas disease: only children can be treated since no effective medicine exists for adults.

A seriously disabling or life-threatening disease can be considered neglected when treatment options are inadequate or do not exist, and when their medicine market potential is insufficient to attract investment by the private sector. A distinction between "*neglected*" and "*most neglected*" can also be made.

For the "most neglected" diseases, patients are so poor that they have virtually no purchasing power, and market forces are too insignificant to stimulate interest among pharmaceutical companies. Examples of "neglected diseases" include malaria, tuberculosis, human African trypanosomiasis (sleeping sickness), South American trypanosomiasis (Chagas disease), Buruli ulcer, dengue fever, leishmaniasis, leprosy, lymphatic filariasis and schistosomiasis. All but malaria and tuberculosis can be considered "most neglected diseases". Typically, neglected diseases are tropical diseases.

Despite progress made in both the basic knowledge of many infectious diseases and the process of medicine discovery and development, tropical infections continue to cause significant morbidity and mortality, mainly in the developing world. The burden of infectious diseases has been compounded by the re-emergence of diseases such as tuberculosis, dengue fever, and African trypanosomiasis. These diseases all predominantly affect poor populations in the less developed world.

The discovery and development of most of the medicines currently available to treat tropical diseases was driven by the needs of European colonial services during the early part of the twentieth century. As Western interests drifted away from these regions, tropical diseases have become progressively neglected. An analysis of medicines developed over the past 25 years shows that of the 1,393 new medicines approved and marketed worldwide only 16 (just over 1 per cent) were indicated for tropical diseases and tuberculosis. These diseases affect primarily poor populations and account for 12 per cent of the global disease burden. In comparison 179 new medicines were developed for cardiovascular diseases, which represent 11 per cent of the global disease burden.

2.3.2 Factors leading to neglected diseases

Three key factors have been identified that can collectively contribute to the burden of illness associated with infectious diseases. These are:

· failure to use existing treatments effectively
· inadequate or non-existing treatments
· insufficient knowledge of the diseases.

We might expect that health research would concentrate on areas of greatest need. However, the reality is quite different. Only 10 per cent of the global health research expenditure is devoted to conditions that account for 90 per cent of the global disease burden. This imbalance has been called the *10/90 research gap*.

Lack of effective treatment is largely a consequence of lack of sufficient research and development (R&D) for these diseases. Most R&D on new medicines is financed by and undertaken by the research-based pharmaceutical industry: it argues that R&D is too costly and risky to invest in low-return neglected diseases of the poor. So neglected diseases are unattractive targets for commercial R&D. They may pose challenges to pharmaceutical science, yet it is a matter of simple economics: potential returns on investment, not public health needs, determine how private companies allocate R&D funding.

Nevertheless, lack of scientific knowledge and economic interest are not the major barriers to medicine development. Existing treatments for infectious diseases are increasingly ineffective due to poor diagnostic options, growing drug resistance, lack of affordability, poor distribution, and inadequate health systems. Policy issues are a major obstacle to the translation of this knowledge into actual benefit for patients. The neglect of development of new medicines for diseases of poverty is therefore a result of both market failure and of failures in public policy.

2.4   Strategies for developing medicines and ensuring access

So, as we have seen, the failure to address neglected diseases does not rest entirely on the shoulders of the private pharmaceutical industry. Governments hold the ultimate responsibility for ensuring that the basic health needs of people are met. They have to take appropriate action when market forces fail to address these needs. However, the responses of individual governments have often been entirely inadequate. Governments have the power to influence medicine development through both direct research funding (either to the pharmaceutical industry, universities or research councils) and policies to influence the activities of the private sector (such as financial incentives, tax measures or subsidies).

2.4.1   Philanthropy and foundations

It is notable that the most important funding for neglected diseases over the past few years has come not from private industry or the public sector, but from increased interest and commitment from philanthropic individuals and foundations. For example, The Bill and Melinda Gates Foundation, in addition to providing substantial funding for vaccines, has become a major force in neglected disease medicine development. However, whilst additional support from foundations is welcome, foundations cannot and should not take the place of public sector responsibility. Private philanthropy can be neither a substitute nor an alibi for government inaction.

2.4.2   Public-private partnerships

Another type of policy initiative that has become of increasing importance is the public-private partnership (PPP). PPPs attempt to foster R&D for neglected diseases by mobilizing expertise, capacity, and funding from both the public and private sectors. The objective is to develop synergies among public and private stakeholders of pluralistic health systems. Typically, the PPP plays a coordinating and management role around a disease-specific R&D agenda, and seeks a combination of public funding, philanthropic donations and in kind donations from industry. Various initiatives and programmes funded in this way are seeking to redress the R&D imbalance.

Major examples of this approach are the *Medicines for Malaria Venture* (MMV), the *Global Alliance for TB Drug Development* (GATB), and *International AIDS Vaccine Initiative* (IAVI). These have been established to ensure the availability to affected populations of medicines for specific diseases. However, the need for a more integrated approach in tackling these major diseases has been recognised by more recent global partnership initiatives such as *The Global Fund to fight AIDS, Tuberculosis and Malaria*. This has focused on stimulating further interest in financing the provision of medicines for the world's top three killer infectious diseases: HIV/AIDS, malaria, and tuberculosis. We return to these issues in chapter 13.

So far, no PPPs have been designed specifically for developing medicines for the "most neglected diseases".

2.4.3   Not-for-profit initiatives

A not-for-profit initiative to address deficiency of R&D for neglected diseases is the *Drugs for Neglected Diseases initiative* (DNDi), created in 2003. This is a needs-driven global medicine development network with the vision to seek equitable access to medicines and a mission to lead research and development for new medicines for neglected diseases. The DNDi intends to develop new medicines or new formulations of existing medicines for patients suffering from the most neglected communicable diseases.

Acting in the public interest, the DNDi aims to plug the existing R&D gaps in essential medicines for these diseases. It will do this by initiating and coordinating medicines R&D projects in collaboration with the international research community, the public sector, the pharmaceutical industry, and other relevant partners. In pursuing these goals, the DNDi proposes to manage R&D networks built on south-south and north-south collaborations and solidarity. It intends to use and support existing capacity in countries where the diseases are endemic, and contribute to building additional capacity in a sustainable manner through technology transfer in the field of medicines R&D for neglected diseases.

In conclusion, more comprehensive global solutions involving all stakeholders concerned are needed to address the R&D crisis in a sustainable way. General recommendations for tackling the issue of neglected diseases are given in Box 2.6.

2.5   Conclusion

We live in a world of enormous inequalities, and nowhere is this more so than in access to health care, treatment and medicines. Whilst the number of medicines on the world market has increased immensely in the last century, millions of people worldwide still do not have access to essential medicines that are affordable and of good quality. Access to medicines means access to treatment. Improving access to quality treatment is currently the most important strategy to reduce death and dis-

---

Box 2.6: Action plan for neglected diseases

· Reorient and redefine a needs-driven research agenda at global level.
· Create north-south and south-south partnerships.
· Involve developing countries as key players in priority setting.
· Develop strong research coalitions.
· Encourage governments to proactively assume responsibility for and to compensate for private market failure.
· Increase long term funding for research of neglected diseases.
· Build R&D capacity in developing countries.
· Transfer technology to increase R&D expertise and infrastructure.
· Promote public-private partnerships (PPP).

Source: DNDi (2003) http://www.dndi.org

---

ability from many diseases. More generally, ensuring access to effective treatment is a high priority issue for international public health.

Medicine accessibility, availability, affordability and acceptability are factors determining whether people will get treatment and care. Various strategies have been proposed and embraced, and the call for treatment access through campaigns and initiatives has had a significant impact. Constructive and multiple solutions are needed that can reduce the inequities in access to medicines whilst at the same time protecting the incentives for research and development. Many steps have already been taken to improve access to medicines and treatment in response to this bleak situation.

Despite impressive advances in science and medicine, society has failed to allocate sufficient resources to fight the battle against those diseases that particularly affect people in poor countries. The lack of R&D for neglected and most neglected diseases means that all too often health staff in developing countries do not have effective medicines to treat many of the diseases they see every day.

However, encouraging initiatives have emerged to counter the market and public policy failures that have led to this crisis. Many of these initiatives are promising and offer real hope for the future. The implementation of new solutions, such as the not-for-profit initiative for developing medicines for neglected diseases, will be essential.

Further reading

Essential Drugs Monitor 28 (2000) Geneva: World Health Organization.
Essential Drugs Monitor 29 (2000) Geneva: World Health Organization.

"Fatal imbalance: The crisis in research and development for drugs for neglected diseases" (2001) *Médecins Sans Frontières Access to Essential Medicines Campaign and the Drugs for Neglected Diseases Working Group*. Paris: Médecins Sans Frontières.

Trouiller, P., Olliaro, P., Torreele, E., Orbinski, J., Laing, R. and Ford, N. (2002) "Drug development for neglected diseases: a deficient market and a public-health policy failure". *Lancet* 359: 2188–2194.

Reich, M. (2000) "The global drug gap", *Science* 287: March 1979.

Sachs, J. et al. (2001) *Macroeconomics and Health: Investing in Health for Economic Development*. Geneva: World Health Organization.

Chapter 3
Assessing the Pharmaceutical Needs of Patients and Populations

Reinhard Huss

Box 3.1: Learning objectives for chapter 3

By the end of this chapter you should be able to:

· Describe incidence of diseases as a method to determine medicine needs of populations.
· List different perspectives on medicine needs.
· Describe methods to determine priorities.
· Explain the meaning of culture and its influence on medicine needs.
· Describe the role of pharmaceutical anthropology in the description and understanding of medicine needs.
· Describe the patient-as-consumer approach to medicine needs.
· Define the terms compliance, adherence, concordance and empowerment.
· Explain the difference between these terms.

3.1   Introduction

This chapter looks at the factors that influence the pharmaceutical needs of both individual patients and whole populations. Medicine needs as expressed by patients may be influenced by a wide range of factors such as the health care system, the financing and pricing of medicines, the pharmaceutical industry and the prevailing local culture. The development of standard treatment guidelines and the calculation of medicine needs should therefore be based on the local incidence of health problems.

In assessing need, we must first consider the cultural construction of health problems, the local responses to them, and the meaning and importance of available medicines to the local population. This chapter describes the methods available to us for assessing the incidence of disease; it explores the impact of culture on the use of medicines; it considers the relevance and value of pharmaceutical anthropology; it explores the role of patients as consumers; and it describes the issues of compliance, adherence, concordance and empowerment in relation to medicines. We begin, however, by taking a population perspective.

3.2   Public health pharmacology

Taking a population perspective to the availability and use of medicines has been referred to as public health pharmacology. The aim is to achieve the best use of available resources in order to advance the universal human right to adequate health care. The taking of a population perspective to pharmaceutical services is referred to as pharmaceutical public health (see also chapter 4). This perspective of public and professional needs may be in conflict with the needs of individual patients. Some patients will be able to afford high prices for medicines; others will have a belief system that results in a preference for ineffective medicines.

Such different perspectives of patients and populations can be examined either from a universal biomedical or from a local indigenous viewpoint. But effective medical treatment requires the active cooperation of both health professional and patient, and both perspectives are needed.

Medicines are not only chemical products that have the capacity to treat diseases: they also facilitate social and symbolic processes amongst people, which influence their attractiveness in different cultures. This may be linked to the shape, colour, form or preparation of the pharmaceutical product. The demand for injections in certain cultural contexts has been well documented, whilst analgesic suppositories are popular in Germany but disliked by British patients.

Medicines as commodities can be passed from one cultural context to another. They can be treated in many different ways: on the one hand they may be considered as ordinary commercial goods, whilst on the other people may endow them with the power to prevent disease, cure illness and alleviate complaints. Health professionals using medicines need to be aware of these factors if they wish to establish an equal and effective therapeutic partnership between themselves and patients.

3.3   Incidence of disease

Information about the incidence of diseases in an area or a country is the essential foundation necessary for planning purposes. Information is necessary in order to:

- · develop standard treatment guidelines,
- · determine the pharmaceutical needs of the population, and
- · estimate the budgetary requirements for these needs.

The incidence of diseases can be calculated according to a formula, as follows:

$$\text{Incidence rate} = \frac{\text{Number of new cases of a disease}}{\text{Population at risk}} \text{ over a period of time}$$

To determine the incidence rate a population at risk has to be followed prospectively over a period of time, and the number of new cases has to be recorded. The diffi-

Box 3.2: Challenges in measuring incidence rate of diseases

Requirements	Challenge
1. Health status of the population	Adequate information about the health status of the population is necessary to decide whether a person is diseased. This may be obtained from health records in health facilities or national health information systems, but it may require a more systematic examination of the population. The latter is necessary if many diseased people either do not attend the modern health facilities or use non-biomedical facilities.
2. Time of onset	With certain health problems and chronic diseases the onset may be indefinite. In these cases an operational date of onset has to be defined.
3. Specification of numerator	If the incidence is calculated for a one-year period, a person may have several disease events within one year, e.g. acute gastro-enteritis. For the purpose of medicine requirements we want to calculate the number of disease events and not the smaller number of diseased people.
4. Specification of the denominator	The denominator should only include the people at risk in a population. For instance men cannot suffer from vaginal discharge. In this example the calculation may be easy. It is more difficult to determine when children with protective antibodies are not susceptible to tetanus.
5. Period of observation	The occurrence of a health problem or disease may vary over time. This can be seasonal within a year, cyclical over several years or secular over longer periods.

culties arise from the gaps in the health information system of a country, and the transformation of the available information into relevant knowledge. Some of the requirements and challenges of assessing the incidence of disease are listed in Box 3.2.

The *morbidity method* and some of its difficulties are listed in Box 3.3. Whilst the scientific soundness of the method is evident for the development of standard treatment guidelines, the calculation of medicine needs works best with a limited number of health facilities or specific health care services.

There are several problems associated with the morbidity method. Morbidity data may have only been collected for a limited number of diseases, or the data may be incomplete or of poor quality. Moreover many health problems in primary health care settings are not well-defined diseases, so that they may not be reported in a uniform and standardised way in health care statistics. Information gaps about the incidence of health problems are not the only limiting factor. Others include insufficient

Box 3.3: The morbidity method and its difficulties

Information needed	Problems
1. Population ⇓	Population figures in low-income countries are not always reliable and may vary considerably from different sources.
2. Coverage ⇓	The percentage of the population which is served by the health care sector depends on geographical, financial and cultural accessibility. These factors may vary over shorter or longer periods according to changes in the sector.
3. Incidence of health problems ⇓	The number of health problems treated in the biomedical sector may depend on and vary with the health professionals providing services, the type of health problems, the health-care seeking behaviour and changes of the population.
4. Medicine needs according to standard treatment guidelines (STG) ⇓	There may be two or more standard treatments so that a frequency of use has to be estimated for each therapy. Health professionals may not adhere to the STGs for various reasons.
5. Medicine quantities required for the supply system ⇓	Additional quantities of medicine are required to cover theft, wastage and the time interval (lead time) between ordering and receiving new medicines. The latter may vary depending on currency exchange regulations, transport situation and availability of medicines with the suppliers.
6. Costs of medicines	This may vary with changes of unit prices of suppliers, exchange rates, transport costs, custom and tax duties.

knowledge about the health-seeking behaviour of people and the treatment-providing behaviour of health professionals. A good example of this is the introduction of sildenafil in the United Kingdom, and the apparent rise in the incidence of erectile dysfunction. This is illustrated in Box 3.4.

These difficulties highlight some important dilemmas in the decision-making process about the needs and priorities of pharmaceutical services for a population.

· Who should take these decisions and how should they be taken?
· Should public health pharmacology limit itself to a biomedical analysis of health problems?
· Should available resources (in terms of personnel, facilities, equipment and finances and the subsequent development of relevant standard treatment guidelines) be analysed?
· Should an anthropological and sociological perspective be included in the process?

Box 3.4: Impact of introduction of a medicine on incidence of disease

Annual incidence of erectile dysfunction among men aged 40-79, before and after introduction of sildenafil (Viagra) in 1998 in the United Kingdom

Year	Number of cases per 1,000 man years	Year	Number of cases per 1,000 man years
1990	2.60	1996	4.15
1991	2.75	1997	4.15
1992	2.60	1998	6.30
1993	3.45	1999	9.50
1994	3.20	2000	8.65
1995	3.40		

Source: Kaye, J.A. and Jick, H. (2003) "Incidence of erectile dysfunction and characteristics of patients before and after the introduction of sildenafil in the United Kingdom: Cross sectional study with comparison patients". *British Medical Journal* 326: 424–425.

3.4   Needs and priorities for medicines

The tension between these different perspectives is reflected in the question as to whether needs and priorities should be defined from a professional or a user perspective? The *normative needs* defined by health professionals may not necessarily correspond with the *felt needs* of the population. Moreover, some of the felt needs may not be expressed as demand towards the biomedical health sector. A number of factors influence this:

· Individuals will have their own preferences regarding the medicines they wish to obtain from the modern health care sector.
· The need for complete or partial payment may limit access to medicines.
· Patients may have alternative sources for specific treatments such as the traditional health care sector.

Equally, normatively defined needs may not be translated into an adequate supply of medicines for the public sector. This may be due to decisions of the government or health service managers who have set different priorities, or simply be due to managerial and organizational problems that prevent an adequate pharmaceutical supply to health facilities.

Box 3.5 illustrates a number of approaches that can be used to identify how the gap between felt and normative needs, and between the demand for and supply of medicines, can be closed. These are not exhaustive, and their effectiveness needs to

Box 3.5:   Relationship and dynamics between felt and normative needs, demand and supply of medicines

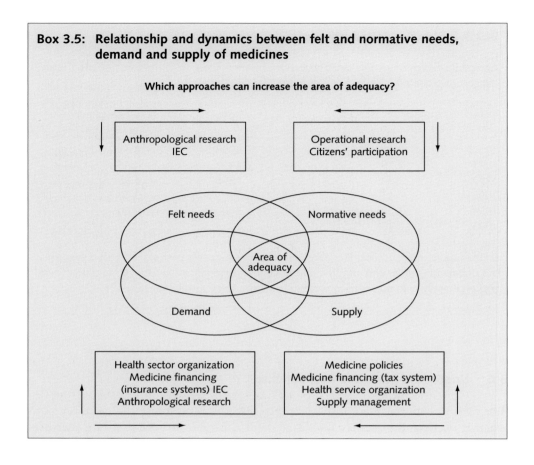

be assessed within a specific health system. *Information, education and communication (IEC)* (see chapter 11) can be an important tool to reduce the gap between felt and normative needs and between demand and supply. Citizens' participation through concerned and organised groups, in such areas as HIV/AIDS, is another important approach to reducing this gap.

3.4.1   Quantitative methods

So far, the question of needs and priorities has focused on the *qualitative* aspect of decision-making, and how the different perspectives and interests can be brought closer together. This section describes the *quantitative* methods that are available to us to calculate the supply needs of a health facility, a district, a province or a whole country, and two methods to set priorities for the supplies in case of insufficient financial means.

Box 3.6: Methods of quantification

Method	Description
Consumption	The past consumption of individual medicines is used to estimate the future need. This estimation has to be adjusted for stock-outs and proposed changes of medicine use.
Morbidity	The estimation is based on the incidence of health problems, the expected frequency of attendance in biomedical facilities and the standard treatment guidelines for these problems.
Adjusted consumption	Data either based on consumption or morbidity method from another geographical area or facility is used extrapolate on the needs in the target system.
Service-level projection of budget requirements	This method estimates medicine costs but not quantities of specific medicines. It uses the average medicine costs per attendance or bed-day in a system of health facilities to estimate the costs in the target system.

These estimations of medicine needs, based on a biomedical and economic perspective, can be described as the *morbidity* and *consumption approaches*. Such a formal quantification of medicines should be mandatory for all initial supplies. Thereafter it should be undertaken on a regular basis in order to avoid stock outs, emergency purchases, overstocks, and to maximize the impact of procurement funds. Based on these approaches we can differentiate between four methods of quantification. These are:

1. Consumption method
2. Morbidity method
3. Adjusted consumption method
4. Service-level projection of budget requirements.

Box 3.6 gives a brief description of each method. Any quantification of medicine requirements may use one or a combination of these four standard methods depending on the disease and the availability of resources and information. The process is inherently imprecise whatever methods are selected, because many variables are involved. A comparison of the methods is given in Box 3.7.

3.4.2   Methods for prioritisation-VEN and ABC

Health service managers have several methods they can call on to prioritize medicine requirements. *VEN* and *ABC analysis* are both frequently used. *VEN analysis*

Box 3.7: Comparison of quantification methods

Method	Uses	Essential Data	Limitations
Consumption	First choice for procurement forecasts, given reliable data. Most reliable predictor of future consumption.	Reliable inventory records. Records of supplier lead time. Projected medicine costs.	Must have accurate consumption data. Can perpetuate irrational use.
Morbidity	Estimating need in new programs or disaster assistance. Comparing use with theoretical needs. Developing and justifying budgets.	Data on population and patient attendances. Actual or projected incidence of health problems. Standard treatments (ideal, actual). Projected medicine costs.	Morbidity data not available for all diseases. Standard treatments may not really be used.
Adjusted consumption	Procurement forecasting when other methods unreliable. Comparing use with other supply system.	Comparison area or system with good per capita data on consumption, patient attendance, service levels, and morbidity. Number of local health facilities by category. Estimation of local user population broken down by age.	Questionable comparability of patient populations, morbidity, and treatment practices.
Service-level projection of budget requirements	Estimating budget needs.	Utilization by service levels and facility type. Average medicine cost per attendance.	Variable facility use, attendance, treatment patterns, supply system efficiency.

Source: Adapted from Quick, J.D., Rankin, J.R., Laing, R.O., O'Connor, R.W., Hogerzeil, H.V., Dukes, M.N.G. and Garnett, A. (eds). (1997) *Managing Drug Supply: The Selection, Procurement, Distribution and Use of Pharmaceuticals*. Second edition. West Hartford, CT, USA: Kumarian Press.

takes a health perspective, whilst ABC analysis (also known as Pareto or 80/20 analysis) takes a cost perspective. Both methods can be used to set priorities in the areas of selection, procurement, distribution and use of medicines.

The *VEN method* was developed in Sri Lanka and sets priorities according to the health impact of medicines. They are divided into three categories: *V* for vital; *E* for essential; and *N* for non-essential. V medicines are potentially life-saving, E medi-

Box 3.8: Criteria for VEN categories

Characteristic of medicines or health problems	Vital	Essential	Non-essential
Frequency of health problems			
- People affected (as % of population)	over 5	1–5	less than 1
- People treated (per day in average health centre)	over 5	1–5	less than 1
Severity of health problems			
- Life-threatening	yes	sometimes	rarely
- Disabling	yes	sometimes	rarely
Effect of medicine			
- Proven efficacy	always	usually	sometimes
- Prevention of serious diseases	yes	no	no
- Cure of serious diseases	yes	yes	no
- Treatment of minor problems	no	sometimes	yes

Source: Adapted from Quick, J.D., Rankin, J.R., Laing, R.O., O'Connor, R.W., Hogerzeil, H.V., Dukes, M.N.G. and Garnett, A. (eds). (1997) *Managing Drug Supply: The Selection, Procurement, Distribution and Use of Pharmaceuticals*. Second edition. West Hartford, CT, USA: Kumarian Press.

cines are effective against serious diseases, whilst N medicines are used for minor and self-limiting health problems. In order to gain wide acceptance the categorization process should be transparent and be based on comprehensive and comprehensible criteria. These are listed in Box 3.8.

The *ABC method* is also called the *80/20, or Pareto analysis*, named after an Italian economist. Pareto described the fact that in many European countries about 80 per cent of the wealth is owned by about 20 per cent of the population. It has since been discovered that this 80/20 rule can be applied to many other areas including inventory and quality management. So, for example, about 20 per cent of all pharmaceutical items are responsible for about 80 per cent of the costs in an organization.

The ABC method is illustrated in Box 3.9. It combines price and volume data. It can be used to categorize medicines into *A, B and C items*. *A* items may be high cost plus high to moderate volume, or high volume plus high to moderate cost items: they are major contributors to purchasing costs. *B* items are moderate contributors and *C* items are minor contributors to purchasing costs.

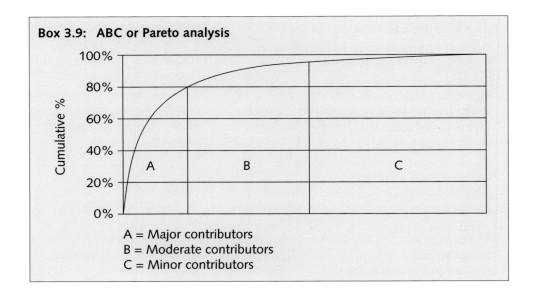

Box 3.9: ABC or Pareto analysis

A = Major contributors
B = Moderate contributors
C = Minor contributors

3.5   Culture and the use of medicines

Information about the incidence of diseases is an important prerequisite for the development of treatment guidelines and the quantification of medicine requirements. However, other factors need to be taken into account. Some medicines, such as analgesics, may be used for multiple conditions: the incidence of diseases and the theoretical quantification of medicine needs tells us nothing about the beliefs that people hold about medicines, which are probably the greatest influence affecting medicine taking. These beliefs are often at variance with the best evidence from medical science, and consequently receive scant attention. Yet they are firmly rooted in personal, family and cultural experiences.

3.5.1 Classifying culture-ETIC and EMIC

Whilst the phrase "personal and family experience" describes a vast but well-defined area, the term "culture" may have several different meanings. The *EMIC* framework for cultural studies of illness is probably the most appropriate approach in relation to medicines. It is concerned with locally valid representations of illness. It includes the local experience, the meanings attached to illness, and risk-related and help-seeking behaviours relating to illness. This perspective is important for health professionals who wish to implement health or medicine policies in a local context. In these circumstances an understanding of the local communities, and of the locally specific features of any general policy, are essential.

The other perspective is called *ETIC* and is concerned with disease rather than illness. It is based on the findings of academic research into health problems and

---

Box 3.10: ETIC and EMIC perspectives of health problems

ETIC – Professional, Universal	EMIC – Local
Disease	Illness
Aetiology – Risk factors	Meaning – Perceived causes
Pathophysiology – Symptoms and signs	Experience – Patterns of distress
Diagnostics, therapeutics – Care providing	Behaviour – Help seeking

---

their treatment. The terms EMIC and ETIC originate from linguistic research, where EMIC refers to the insiders' perspective and ETIC to the outsiders' perspective. The main features of the two approaches are summarised in Box 3.10.

3.5.2 Classifying culture-group and grid

The classification of cultures according to the criteria of *group and grid* may also be relevant for the improvement of pharmacotherapy. The *group* criterion describes who decides and acts, whilst the *grid* criterion defines how social interactions come about. They are summarised in Box 3.11.

In high group plus high grid cultures, which are often associated with traditional societies, the use of medicines is determined more by the group experience and the interaction between the social and natural worlds. Modern society is often described as a low group plus low grid culture. However, the increasing popularity of alternative healing methods, and the recognition of social stress in industrialized societies, is evidence that reality is rather different.

The selection and use of medicines should therefore take into account not only the local epidemiological disease patterns, but also the contributions of social and cultural epidemiology. The social causes of diseases, as well as their local interpretation, have to be understood in order to promote the rational use of medicines. Otherwise the marketing and use of medicines may not only fail to meet the needs of the target population but also elevate the health risks to which its members are exposed.

Box 3.11: Group and grid dimensions of culture

Group	high	Natural and social world are completely integrated
	low	Individual, society and nature are seen as separate
Grid	high	Social relations are rigidly circumscribed
	low	Spontaneity, flexibility and variability are allowed

Source: Douglas, M. (1973) *Natural Symbols: Explorations in Cosmology*. New York: Vintage Books.

To better address the needs of the local population in its cultural context, the expressed medicine needs have to be analysed with a culturally sensitive methodology, since medicines are much more than just "pills that contain healing powers". Treatment guidelines to improve the health services for a population can only be developed when there is a detailed knowledge of the beliefs of medicine taking by the people concerned.

3.5.3   Exploring culture-pharmaceutical anthropology

Pharmaceutical anthropology is a sub-discipline of anthropology that studies pharmaceutical-related behaviour and the local interpretation of medicines in different social settings. The EMIC and ETIC perspectives of cultural epidemiology provide complementary information, so that health care organizations are better informed. This will assist them in developing locally appropriate and adapted treatment guidelines, capable of providing health professionals with important guidance on how to use medicines.

Whilst in most Western societies it is mainly the individual who decides about medicine use, in many societies around the world therapy is embedded in kinship and community relationships. Decisions are often made by a wider therapy management group in the community rather than by the individual. In this respect periods of illness are occasions of dependency and social control. They provide an opportunity to review social relationships and conceptions of the person in the world.

3.5.4   The status of medicines in society

The availability of medicines at local markets in low-income countries enables the individual, often for the first time, to manage the illness personally without having to involve the family, kin, neighbours, traditional or even biomedical specialists. These medicines are almost always sold in units of a single pill, as people are often not able to afford more. In this way social control can be evaded and illness symptoms may be kept hidden from the family or the community for a considerable length of time.

Modern medicines transform dependent persons into responsible human beings who take their health into their own hands. Biomedical pharmaceuticals are usually ready to be taken: they are not normally the products of a long journey involving elaborate and special preparation, as is often the case for herbal or traditional treatments. Medicines are tangible items or commodities that can be exchanged for money, and they possess characteristics that may simultaneously be private, secret, liberating and convenient. For example, the taking of the oral contraceptive pill is often the subject of much secrecy and subterfuge.

The expressed medicine needs of patients demonstrate that modern medicines are endowed with properties that extend far beyond their role as powerful therapeutic agents. It has been noted that medicines are interpreted in relation to their physical properties (e.g. colour, taste, appearance), as well as their form (e.g. liquid, tablet, capsule, injection). Hausa considered bitter medicines to be dangerous for pregnant women because of possible effects on the uterus and gastrointestinal tract: women often used these "secondary" effects when abortion, for instance, was in fact the desired outcome.

Taste and colour play an important role in the course of medicine taking, as do their expected side effects. The interpretation of signs or symptoms is deeply embedded in cultural meanings of therapeutics and outcome, and patients and healers in the same society do not necessarily agree on what is the "primary" effect of a medicine and what is "secondary" to effective therapy. Another example is the use of diuretics when patients wish to lose weight.

3.5.5   Differing perceptions of medicines

The cultural meaning of a health problem, the applied therapeutics and the expected outcome all influence the interpretation of signs or symptoms of what is accepted as a successful therapy. Patients and practitioners may hold different viewpoints about primary and secondary effects of an effective therapy. For instance patients may expect or take medications to first induce vomiting or diarrhoea in order to clean the gastrointestinal tract and then use additional treatment to heal the digestive system. The different viewpoints and beliefs result in a gap in understanding and communication between the patient and the prescriber.

The medicine needs of a population should not therefore be analysed in isolation from their cultural construction. Patients use medicines as independent items that do not require the control of practitioners or family. Other treatments (for example surgery) cannot be isolated from the professional control of the surgeon.

At the national level these investigations have to take account of research on the decisions and practices of other key stakeholders, such as the multi-national pharmaceutical industry. Political and economic factors also influence global as well as national availability, marketing and distribution of medicines.

Finally, the development of an anthropology of pharmaceutical practice requires socially informed health service research and pharmacoepidemiological studies, about the impact of medicine prescription, dispensing and consumption patterns on the management of diseases, drug resistance and iatrogenic health problems. We return to such studies in chapter 12.

3.6   Patients as consumers

The motivation to obtain and use medicines is not simply that they are powerful, but also that people believe them to be so. People take medicines in front of significant other people to reinforce the declaration of their health status, or possession of a special condition. Medicines are exchanged in communities and families as symbols of care to reduce suffering and foster hope. In this sense, medicines serve as non-verbal communicators of life problems.

Certain medicines, such as vitamins, are seen to possess power and prestige. They are welcome as commodities in many traditional societies, although they might not actually be needed for treatment. This phenomenon also occurs in developed societies, where the consumption of powerful medicines appears to be the answer for many challenges and difficulties in life. The boundaries of treatable conditions are expanding fast: they now include natural processes such as normal ageing, personal characteristics such as shyness and baldness, and any form of human distress.

3.6.1   Direct to consumer advertising

The marketing of pharmaceutical companies often creates and reinforces the belief that "there is a pill for every ill" and "the more pills the better the treatment'. Indeed, critics of the industry claim that the ultimate aim of the industry is the corporate construction of illness and treatment, and the transformation of all healthy citizens into patients, purchasers and consumers of pharmaceutical products.

Companies market brand name medicines like lifestyle products, and lobby for a decrease in the regulation of the industry by allowing *direct-to-consumer advertising (DTCA)* of prescription medicines. This is currently only permitted in two industrialized countries, the United States and New Zealand. However, DTCA has already been reported to have negative effects on public health goals: it has created difficulties in ensuring adequate funding and equity of services, and also in patient safety. DTCA has in addition been shown to undermine the therapeutic relationship between patient and prescriber.

Persistent demands from the pharmaceutical industry for DTCA to be allowed in all countries is seen by some analysts as confirming the predictions of a number of social scientists about the consequences of an increasing automation of production. They have suggested that the most successful companies will ultimately control the consumption rather than the production of goods, and that DTCA is evidence of this in relation to the pharmaceutical industry.

3.6.2   Consumer sovereignty

In taking a public health pharmacology perspective, we need to analyse critically the ambivalent social role of patients as consumers. Consumer and self-help organizations now promote the autonomy of patients, who are encouraged to inform themselves and trust their own judgement regarding treatment decisions. This is a welcome change from the former situation in which adult patients were treated more like children or other dependents. However, a situation of complete sovereignty of patients as consumers may undermine the expectations of citizens to pharmaceutical services which are not only beneficial to some but to all members of society.

This social dilemma can be illustrated by means of the *diamond and water paradox* described by economists. Water is an essential factor for human life, whilst diamonds are not. Nevertheless, diamonds are far more expensive and lucrative to sell in society than water. The high price of diamonds is explained by the fact that, despite their limited purposes, people still want them. Unlike water, diamonds are rare. But they would soon lose their high price if they became as readily available as water, a fact well understood by the diamond producing industry: the major producer ensures a quasi monopoly in order to maintain very high prices.

The aim of public health pharmacology is to secure the global right of all people to the pharmaceutical equivalent of water. Therefore essential medicines have to be cheap and accessible for everybody, just like water. This essential human right of all citizens to basic and adequate pharmaceutical services needs to be protected and given social priority over the demand of some consumers to special pharmaceutical products of limited or doubtful use.

3.7 Compliance, adherence and concordance

An integral part of patients' beliefs about medicines relates to how they are taken. Patients may be more or less likely to take medicines depending on such factors as who prescribed them and how serious their condition is. Three terms are used in this connection: compliance, adherence and concordance. All three are concerned with patient's behaviour in relation to prescribed medicines, along with empowerment. We will consider each of them in turn.

3.7.1 Compliance

For a long time the term *"compliance"* has been widely used to describe the extent to which the behaviour of a patient coincides with the corresponding clinical prescription. The term has been qualified as "sufficient compliance to achieve the desired clinical outcomes'. This takes into account the fact that some therapeutic regimens may allow more flexibility than others. Subsequently, there has been a change from the term "compliance" to "adherence". This is more than simply a change of words, because it represents a different understanding of the relationship between the patient and the prescriber.

3.7.2 Adherence

Whilst compliance assumes a passive acceptance of the prescriber's instructions, *"adherence"* underlines the importance of the attitude towards the patient and the communication skills of the health care providers. Unfortunately both are often neglected in the training of providers. Non-adherence rates of more than 50 per cent

---

Box 3.12: Some causes of non-adherence to prescribed treatment by patients

---

· Inappropriate attitudes of health care providers
· Poor communication skills of providers
· Patients' fear of asking questions
· Inadequate consulting time
· Inadequate dispensing time
· No or inappropriate printed information for patients
· Inability to pay for prescribed medicines
· Fear of adverse drug reactions
· Adverse drug reactions
· Improvement of health problem or disease
· Advice from relatives or members of the community
· Social and professional situation
· Disabilities of patients
· Complexity and duration of treatment, particularly in chronic diseases

---

have been reported. Some of the possible causes of non-adherence to prescribed treatment are listed in Box 3.12.

3.7.3 Concordance

A further step in the direction of seeing patients as equals is the term *"concordance"*. This is based on recognition that health care providers and patients often hold different health beliefs. The former are mostly trained in the biomedical model, comprising scientific evidence and technical expertise that is derived from population-based studies in a well-defined and controlled environment.

Patients' health beliefs originate from their experience and personalities and the family and the culture in which they live. The optimal use of medicine may have been defined in a well- controlled scientific environment, but this evidence has to be interpreted in the "real" context of the individual patient. However, "reality" is a complex environment where scientific predictability exists alongside (and is sometimes replaced by) high levels of uncertainty.

The term "concordance" describes a process between equal partners where both sides describe and discuss their intentions during their clinical encounter. These negotiations should assist the patient in making as informed a choice as possible about the treatment. The aim is to reach an understanding and establish a therapeutic alliance between prescriber and patient.

This approach encourages a fundamental shift in power between the two parties. It recognizes that patients carry out experiments on themselves in order to achieve the desired and promised health improvements. If a patient and prescriber fail to achieve this understanding on the therapeutic process, benefits and risks, the consultation has failed and can be described as non-concordant.

3.8 Empowerment

The ultimate aim of pharmaceutical equality should be individual and collective empowerment. Power can be defined as "the ability to determine the interests of people". In this respect a clinical encounter is a specific power situation where medical practitioners may exercise enormous power over the users of their services. This may be unavoidable in an emergency, when a person is seriously ill. In this case users have little option but to entrust the power to determine their best interests to the health care provider.

In this situation the patient places enormous trust in the health care provider, and the provider exercises enormous power over the patient. However, today such a situation should be limited to extreme and special circumstances where no alternative approach is possible. Illich has used the term *nocebo effect*, meaning a negative placebo effect, when health care services transform users into passive beings who are unable to mobilize their own self-healing resources.

Primary health care, where most therapeutic encounters occur, should have a vision that therapeutic partnership encourages self-determination. Under these circumstances users are able to define their own interests with respect to pharmaceutical treatment. This is not only a question of human dignity but also of realism, as about 80 per cent of all medications in low income countries are sold over the counter. This is itself a strong argument in favour of people being made more knowledgeable about the risks and benefits of frequently used medicines.

3.8.1 Individual risk assessment

There are two types of inter-related risks and benefits, personal and social, which have to be considered when people are given increased freedom to determine their own treatment. The decision as to whether a particular medicine should be made available only on prescription or over the counter needs to be taken on the basis of a risk assessment for individuals and societies. Questions that need to be addressed include:

· Does the prescription procedure protect individuals because of their limited knowledge from the risks of this treatment?
· What are the risks of avoidable morbidity and mortality in the population of a specific country, resulting from the treatment being delayed or withheld because of the prescription procedure in the health care system?

A few examples illustrate the choices to be made. Paracetamol can be legally obtained over the counter in many countries. This situation is the result of just such a risk assessment, which has taken account of all the relevant facts, including the fact that it is possible for someone to take their own life with an overdose of paracetamol.

In Zimbabwe the first line treatment of chloroquine against malaria was at one time available in many rural shops, because it was judged that the population had sufficient knowledge about this treatment. In addition, the epidemiological risk of severe malaria due to delayed prescriptions was judged to be higher than the risk of inappropriate use of chloroquine among the rural population.

It has been argued that oral contraceptives should be available over the counter in many countries where maternal mortality is very high. This is because the health benefits of freely available oral contraceptives are almost always higher than the risks of their inappropriate use, in view of the morbidity and mortality linked to normal pregnancies in these countries.

3.8.2 Social risk assessment

In addition to the individual risk assessment there needs to be a social risk assessment. The inappropriate use of certain medicines does not only carry risks for the individual but also for society. These include the development of resistance to antibiotics and the abuse of medicines. The development of microbial resistance to certain medicines can incur enormous social risks and costs. Equally the misuse of sedatives can contribute to social problems and costs. Strict prescribing and dispensing procedures are one way to protect society against these risks, but at the price that individuals may not be able to obtain the treatment they need at the time they need it.

Monopolies such as prescribing and dispensing guaranteed by the state carry inherent risks of abuse for the members of society. For example, medical practitioners may prescribe medicines that are not clinically indicated for personal economic interests, and to ensure that they will not lose their clients. Widespread education about medicines is an alternative means of reducing the risks of inappropriate use and of empowering people to make their own decisions about the best treatment, assisted by the professional advice of health care providers.

3.9   Conclusion

This chapter has been concerned with assessing the medicine needs of patients and populations. We have considered some of the tools and techniques available to us, and explored the impact of culture on perceptions about their use. We have concluded that the most important factor is to empower people to make their own decisions about medicine use.

Pharmaceutical empowerment of populations is intimately related to the regulatory framework for medicines in particular countries. Empowerment can only be achieved by the two coming together; at the same time the scientific perspective of experts can be greatly enhanced by the benefit and risk assessments of citizens. This process is not only an important facet of democracy, but it also takes account of the fact that even expert opinions are always a mixture of knowledge and the values held by these experts. Furthermore, citizen participation in pharmaceutical decision-making offers the opportunity for medicine regulation to become more transparent, acceptable and workable.

Health professionals need to promote the health education of both individuals and societies. This will enable individual treatment to increasingly become the decision of the person concerned, assisted by the professional advice of their health care providers. Societies too need to be empowered to take appropriate decisions that balance the interests of individuals to decide their own treatment and the interests of society to protect the common good for present and future generations.

Further reading

Illich I. (1977) *Limits to Medicine. Medical Nemesis: The Expropriation of Health*. Harmondsworth: Penguin.

Idler, E.L. (1979) "Definition of health and illness and medical sociology". *Social Science and Medicine* 13: 723–731.

Mintzes, B., Barer, M.L., Kravitz, R.L., Bassett, K. et al. (2003) "How does direct-to-consumer advertising (DTCA) affect prescribing? A survey in primary care environments with and without legal DTCA". *Canadian Medical Association Journal* 169 (5): 405–412.

Moynihan, R., Heath, I. and Henry D. (2002) "Selling sickness: the pharmaceutical industry and disease mongering". *British Medical Journal* 324: 886–891.

Quick, J.D., Rankin, J.R., Laing, R.O., O'Connor, R.W., Hogerzeil, H.V., Dukes, M.N.G. and Garnett, A. (eds.) (1997) *Managing Drug Supply: The Selection, Procurement, Distribution and Use of Pharmaceuticals*. Second edition. West Hartford, CT, USA: Kumarian Press.

Nichter, M. and Vuckovic, N. (1994) "Agenda for an Anthropology of Pharmaceutical Practice". *Social Science and Medicine* 39 (11): 1509–1525.

RPSGB (1997) *From Compliance to Concordance: Achieving Shared Goals in Medicine Taking*. London: Royal Pharmaceutical Society of Great Britain.

Sackett, D.L. and Haynes, R.B. (eds.) (1976) *Compliance with Therapeutic Regimens*. Baltimore & London: The Johns Hopkins University Press.

Trostle, J. (1996) "Inappropriate distribution of medicines by professionals in developing countries". *Social Science and Medicine* 42 (8): 1117–1120.

Van der Geest, S. and Whyte, S.R. (1989) "The charm of medicines: metaphors and metonyms". *Medical Anthropology Quarterly* 345–367.

Weiss, M.G. (2001) "Cultural epidemiology: an introduction and overview". *Anthropology and Medicine* 8 (1): 5–30.

WHO/DAP (1988) *Estimating Drug Requirements – A Practical Manual*. WHO/DAP/88.2. Geneva: World Health Organization.

Woloshin, S. et al. (2001) "Direct-to-consumer advertisements for prescription drugs: what are Americans being sold?" *Lancet* 358: 1141–1146.

Chapter 4
The Role of Health Professionals

Karin Wiedenmayer

Box 4.1: Learning objectives for chapter 4

By the end of this chapter you should be able to:

· List a range of health professionals who handle medicines.
· Describe the role of doctors in the management of pharmaceuticals.
· List the steps involved in selecting a P-drug.
· List the information that should be included on all prescriptions and the reasons for it.
· Describe the characteristics of the eight star pharmacist.
· Indicate the levels in health care systems in which pharmacists are involved.
· Describe WHO initiatives concerning the training of pharmacists in low income countries.
· Describe the role of traditional healers in the management of medicines.
· Compare and contrast traditional medicine and complementary and alternative medicine.
· Describe how traditional and western medicine are becoming integrated.

4.1 Introduction

Health care is provided in many different settings and by a variety of health professionals. Medicines are an important component of medical care and treatment, and they are handled by a number of different health professionals. Situations where medicines are transferred and handed over from one person to another have been described as "medicine encounters". Health professionals are usually involved in the prescribing, dispensing, administration and selling of medicines, but they are not always the only ones to undertake such activities.

In many developing countries with weak medicine regulatory mechanisms, medicines are often to be found in markets and shops, or available from medicine peddlers. Health professionals may not be involved at all. Throughout the world medicine encounters occur in families and communities without professional input: with the advent of electronic communication technology, medicines can be bought over the internet, a practice that so far remains unregulated. Thus, in practice, medicine encounters occur not only in clinical settings, but also in community, family, electronic and commercial settings.

4.2   Classifying health professionals

Health care professionals or practitioners include medical practitioners, pharmacists, nurses, dentists and allied health care professionals. The definition and classification of health care professionals varies from country to country. Areas of practice can be the clinical setting, administration, research and academia. Depending on their professional role and status, as well as their training and involvement in health care, health professionals often have different and multiple responsibilities. In some countries, including some in Africa, pharmacists are considered allied or auxiliary health staff.

In the USA, clinical pharmacists have a prominent clinical role and are integrated into the medical team. Nurses are trained in academic institutions in some countries like the USA and thus may proceed to graduate studies and PhD degrees. On the other hand, nurses in Europe usually receive training from institutions with a more vocational focus. Thus a classification of health professionals will always depend on the context.

Medical practitioners either practice in general areas, or specialize further in areas such as nephrology, cardiology and infectious disease. Dentists have a more narrowly defined area of practice. Pharmacists either practice in the supply chain of medicines or are integrated into clinical activities. Nurses tend to care more for the physical and psychosocial needs of patients, and to carry out instructions from medical practitioners regarding patient care.

Allied health professionals (AHPs) include a variety of professions and functions that usually provide health care under the responsibility and direction of a medical practitioner, pharmacist, nurse or dentist. In some countries they are referred to as paramedics or auxiliary health professionals. In some developing countries with a shortage of medical practitioners, medical assistants may be in charge of health facilities and provide primary health care. Allied health professionals may be given limited authority to diagnose and treat minor disease in primary health care. Such individuals include medical assistants and rural medical aids. Some examples of allied health professionals are listed in Box 4.2.

The *health care team* consists of the patient plus all the health care professionals who have responsibility for patient care. The patient is an important member of the team. Members of the team interact by exchanging information and agreeing shared recommendations concerning patient care. The *medical team*, on the other hand, is organized primarily in hospitals, and is responsible for providing health care services and consultations to patients on a particular ward. In teaching hospitals medical teams play an important role in the training of medical students.

4.3   Health professionals and medicine responsibilities

Health care professionals have varying responsibilities in relation to medicines. Medicine-related tasks may include prescribing, dispensing, administration, supply or sale. However, the situation in practice often differs from that defined by legisla-

---

Box 4.2: Examples of allied health professionals

1. Medical practitioner assistant	9. Laboratory technician
2. Medical assistant	10. Occupational therapist
3. Rural medical aid worker	11. Social worker
4. Pharmacy assistant	12. Rehabilitation therapist
5. Dental assistant	13. Physical therapist
6. Nurse assistant	14. Community health worker
7. Midwife	15. Traditional birth attendant
8. Medical technician	16. Injectionist

---

tion: medicines are a lucrative commodity, and they are frequently prescribed, dispensed and sold by a variety of individuals other than health professionals.

In most countries a clear distinction is made between the prescribing and dispensing of medicines, with the functions being undertaken by separate health professionals. The dispensing of medicines has historically been the role of the pharmacist. However, in some countries this activity is sometimes done by dispensing doctors, who operate a small pharmacy in their practice rooms. The combination of these roles is controversial, and may lead to inappropriate prescribing if doctors have incentives to prescribe excessive or expensive medicines.

In developing countries medicines are generally dispensed by nurses, nurse assistants, unqualified clerks and even family members. This arrangement may influence the quality of the medicine dispensed, the counselling provided and the nature of the instructions given. The administration of medicines such as intravenous infusions and injections is traditionally the domain of nurses, but is also carried out by medical practitioners, pharmacists, injectionists and other allied health professionals.

Medicines are often considered as no different to other commodities, and are correspondingly handled in the same way as any other commercial product. However, every effort needs to be made to ensure that only qualified health professionals handle medicines, based on their qualification and function, to avoid unexpected and adverse drug events.

4.3.1   Selection of medicines for formularies and medicine lists

As many as 70 per cent of medicines on the world market are non-essential or duplicates of others that are readily available. Many are derivatives of older medicines and offer only limited advantages, such as improved pharmacokinetics or different adverse effect profiles. These are sometimes referred to as "me-too" medicines. Increasingly, so-called "lifestyle medicines" penetrate the market worldwide. These include medicines for obesity, hair loss, depression and sexual stamina. Other medicines show high toxicity relative to their therapeutic effects. Many new medicines are not relevant to the therapeutic needs of a given population or health priorities of a particular country. Few new medicines represent real therapeutic advances, one notable exception being the ARVs.

The immense number of medicines on the world market has made it impossible for health professionals to keep up to date and to compare alternatives. On the other hand, it has been estimated that most medical practitioners routinely use less than two hundred medicines in everyday practice, and sometimes far fewer. Medicines can provide great health benefits, but their costs and risks are substantial. The rational selection of medicines can have a considerable impact on both the quality and cost of treatment. Medicine selection interventions are very cost-effective.

A list of medicines may be selected for a health facility, the whole public health sector or nationally (see chapter 8). Limited lists such as the Essential Medicines List produced by WHO are usually developed by consensus, and are also increasingly introduced as formularies in industrialized countries. The goal of rational medicine selection is to define what is most needed, and to identify those medicines that offer the best evidence of efficacy and cost-effectiveness for a given therapeutic need. We consider the rationale for the selection of medicine lists in chapter 12.

4.3.2   Tools for the selection of medicines

The field of pharmacotherapy is changing rapidly, with new products and information constantly being introduced. Health care professionals are faced with the constant challenge of new information, which they need to filter, assimilate and utilize to adapt their practice. Whilst medicine therapy may be one of the most cost-effective of therapeutic interventions, the identification of real improvements in the field is often very difficult because of the marketing practices of many pharmaceutical corporations.

It is therefore essential for specialized health professionals to understand and be able to use the tools of "critical appraisal" and "cost-effectiveness analysis" as they evaluate the huge amount of information that reaches them. The techniques used have been incorporated into the emerging disciplines of evidence-based medicine (EBM) and pharmacotherapy, and pharmacoeconomics (see chapter 11).

Evidence based medicine attempts to move practice and prescribing away from an empirical and anecdotal approach to the best possible evidence for the effectiveness of a medicine. What EBM aims to do is to integrate the best research evidence with clinical expertise and patient values. The process applied in assessing clinical evidence is termed "critical appraisal". The most extensive system world-wide for the review of clinical trials is the Cochrane Database of Systematic Reviews.

However, in many cases, particularly in developing countries (but also in some developed countries), practitioners do not have access to "best evidence", because of the circumstances in which they practice, or they may not have the time or skills to assess the evidence that is available. In such cases, an approach often used is to develop specific prescribing guidelines. In this way the number of choices is restricted to those which are expected to produce the best possible results, particularly in resource-scarce environments. The relevant evidence is used to develop treatment guidelines to assist the process of decision-making and to contribute to rational and cost-effective health care.

Pharmacoeconomics is the discipline used when clinical/therapeutic alternatives are evaluated from the economic viewpoint and the most cost-effective intervention is identified. It allows decision-makers to choose amongst alternative therapies and interventions by establishing the differences between them in quantitative (monetary) or qualitative (utility) terms. We consider this further in chapter 12.

4.4 The role of the doctor

In most countries medical practitioners are traditionally responsible for prescribing. Again, however, this may vary depending on the setting, country and legislation. In some areas nurses are permitted to prescribe certain medicines. Similarly, legislation in some countries gives pharmacists the right to prescribe named medicines under specific circumstances. In developing countries AHPs such as medical assistants or community health workers may be permitted within limits of formularies to prescribe medicines in primary health care.

4.4.1 Selection of P-drugs

The WHO "Guide to Good Prescribing" manual formulated the concept of P-drugs. P-drugs are the medicines that medical practitioners or other prescribers have chosen to prescribe regularly, and with which they are familiar. They represent their first choice therapy, along with some alternative choices for given indications.

The concept of the P-drug involves more than just the name of a pharmacological substance; it also includes the dosage form, dosage schedule and duration of treatment, along with information about adverse effects, contraindications, warnings and prescribing for specific patient groups such as pregnant women. P-drugs differ from country to country, and between doctors: they take account of varying availability and cost of medicines, different national formularies and essential medicines lists, medical culture, and individual interpretation of information. However, the principle is universally valid.

P-drugs enable prescribers to avoid repeated searches for a good medicine in daily practice. Furthermore, as P-drugs are used regularly, prescribers get to know both their pharmacological and adverse effects thoroughly, with obvious benefits for the patient.

Choosing a P-drug is a process that can be divided into five steps. These are:

Step 1:   Define the diagnosis;
Step 2:   Specify the therapeutic objective;
Step 3:   Make an inventory of effective groups of medicines;
Step 4:   Choose an effective group according to criteria;
Step 5:   Choose a P-drug.

Box 4.3: Example of selecting a P-drug: angina pectoris

Step 1: Define the diagnosis
> Stable angina pectoris, caused by a partial occlusion of coronary artery

Step 2: Specify therapeutic objective
> Stop an attack as soon as possible
> Reduce myocardial oxygen need by decreasing preload, contractility, heart rate or afterload

Step 3: Make inventory of effective groups
> Nitrates
> Beta-blockers
> Calcium channel blockers

Step 4: Choose a group according to criteria

	Efficacy	Safety	Suitability	Cost
Nitrates (tablet)	+	±	++	+
Beta-blockers (injection)	+	±	–	–
Calcium channel blockers (injection)	+	±	–	–

Step 5: Choose a P-drug

	Efficacy	Safety	Suitability	Cost
Glyceryl trinitrate (tablet)	+	±	+	+
Glyceryl trinitrate (spray)	+	±	(+)	–
Isosorbide dinitrate (tablet)	+	±	+	±
Isosorbide mononitrate (tablet)	+	±	+	±

Conclusion

Active substance:	glyceryl trinitrate
Dosage form:	sublingual tablet 1 mg
Dosage schedule:	1 tablet as needed; second tablet if pain persists
Duration:	length of monitoring interval

Source: Guide to Good Prescribing (1994) WHO/DAP/95.1. Geneva: World Health Organization.

During the process of selecting a P-drug, three criteria should be used: *safety, suitability and cost of treatment*. An example involving the selection of a P-drug for angina pectoris is presented in Box 4.3.

4.4.2   Prescribing of medicines

A prescription is an instruction from a medical practitioner or prescriber to a dispenser of medicines, traditionally a pharmacist. As indicated previously, the prescriber might be an allied health professional, such as a medical assistant, a midwife or a nurse. There is no global standard for prescriptions: every country has its own

---

Box 4.4: Information that should be included on a prescription

---

1. Name, address, telephone of prescriber
2. Date
3. Diagnosis or diagnostic code
4. Generic name of the medicine, strength
5. Dosage form, total amount
6. Label: instructions, warnings
7. Name, address, age of patient
8. Signature or initials of prescriber

---

standards for the minimum information required for a prescription. Each has its own laws and regulations to define which medicines require a prescription and who is entitled to write it. The most important requirement is that the prescription should be clear. It should be legible and indicate precisely what should be given. Few prescriptions are still written in Latin; the local language is preferred.

Box 4.4 summarizes the data that constitute the core requirements for every prescription. Additional information may be added: this might include the type of health insurance the patient has. The diagnosis is essential for clinical input by the pharmacist. The layout of the prescription form and the period of validity may vary between countries. The number of medicines per prescription may be restricted. Some countries require prescriptions for narcotics on a separate form. Hospitals often have their own standard prescription forms.

4.4.3   Electronic prescribing

A recent development is *electronic prescribing*. This innovation offers two main advantages: improved quality of care, and reduced cost. Other possible benefits include a reduction of errors, ready access to medicine information, process efficiency, security and integration of patient records. Guidelines for electronic prescribing have been issued both by the American Medical Association and the International Pharmaceutical Federation. We consider this issue further in chapter 13.

4.5   The role of the pharmacist

The dispensing and sale of medicines has traditionally been the role of the pharmacist. Pharmacy is the profession concerned with therapeutic medicines: it is the profession that historically has had responsibility for the medicine supply chain, and for ensuring safe, effective and rational medicine use. Pharmacists stand at the interface between research and development, manufacturer, prescriber, patient and the medicine itself. The role of the pharmacist takes different forms in various parts of the

---

Box 4.5: The seven star pharmacist

1. Care-giver	5. Life-long learner
2. Decision-maker	6. Teacher
3. Communicator	7. Leader
4. Manager	8. Researcher (additional)

Source: International Pharmaceutical Federation (2003) www.fip.org

---

world. The pharmacist's involvement with pharmaceuticals can be in research and development, formulation, manufacturing, quality assurance, licensing, marketing, distribution, storage, supply, dispensing and monitoring.

Pharmacists practice in a wide variety of settings. These include community pharmacy (in retail premises), hospital pharmacy (in all types of hospital from small local hospitals to large teaching hospitals), the pharmaceutical industry and academia. In addition, pharmacists are involved in health service administration, in research, in international health and in NGOs.

4.5.1   Traditional role of the pharmacist

In the past (and still today in many countries) the role and functions of the pharmacist were clearly identified with that of the product: the role of the pharmacist was that of compounder, supplier and dispenser of medicines.

The role of the pharmacist, however, is undergoing tremendous change. It is facing serious and exciting challenges, both in industrialized and developing countries. Health care delivery and services are continuously changing. Rising costs of health care, constrained and reducing resources, inefficient health care systems, the burden of disease and the changing social, economic and political environment facing most countries, have made the need for change in health care systems imperative. Pharmacy has an important role to play in this process: but to do so the role of the pharmacist needs to be redefined and re-orientated.

4.5.2   The seven star pharmacist

The International Pharmaceutical Federation (FIP) has defined the term *seven star pharmacist*, describing seven essential functions of the pharmacist in a changing health system context. These are listed in Box 4.5. An eighth star for "researcher" should perhaps also be added.

As we have seen, increasingly medicines can be purchased in new settings, and are handled by non-pharmacists. Compounding has been largely replaced by the commercial manufacture of nearly all formulations. Depending on the country and

> **Box 4.6: Value of professional pharmacist services**
>
> "There is clear evidence across a number of different settings for the effectiveness of pharmaceutical care services, continuity of care services post-hospital discharge, pharmacist education services to consumers and pharmacist education services to health practitioners for improving patient outcomes or medication use.
>
> There is more limited evidence, often limited to one or two countries, but still positive evidence for the effectiveness of pharmacist managed clinics, pharmacist review of repeat prescribing and pharmacist participation in therapeutic decision making in improving patient outcomes.
>
> New professional services that have not yet been adequately evaluated include pharmacist administration of vaccines, pharmacist involvement in pre-admission clinics and pharmacist participation in hospital or home services.
>
> Overall, this review demonstrates that there is considerable high quality evidence to support the value of professional pharmacy services in the community setting. Studies evaluating the majority of professional services currently provided by community pharmacists were located and, importantly, demonstrated improvements in outcomes for patients. Improvement in economic analyses is still required. Where the evidence is sound, consideration now needs to be given to implementing these services more broadly within a country's health system."
>
> *Source*: Roughead, L., Semple, S. and Vitry, A. (2003). *The Value of Pharmacist Professional Services in the Community Setting: A Systematic Review of the Literature 1990–2002.* http://www.guild.org.au/public/researchdocs/reportvalueservices.pdf

situation medicines can be found and bought in supermarkets, in drug stores or on the market: they can be ordered by mail order or over the internet; and they are sold by medical practitioners and dispensed by computerized dispensing machines. Under these circumstances it is pertinent to ask:

· Do we still need pharmacists? and
· What is the value of pharmacy services?

A recent review of the value of professional pharmacist services evaluated the strength of the evidence for the effectiveness of such services in terms of consumer outcomes, and where possible, the economic benefits. The key findings of this review, illustrating the value of professional pharmacist services, are summarized in Box 4.6.

4.5.3 Pharmaceutical public health

The main professional roles and areas of practice of the pharmacist can be differentiated into activities of patient care, public health, and supply and dispensing.

Pharmaceutical care and *clinical pharmacy* represent new models of function and activities at patient level, and help to define the developing role of the pharmacist in the health care system. "Clinical pharmacy" has been described as one component of pharmaceutical care, and focuses on pharmacists' involvement with patient care at inpatient level.

Health promotion, disease prevention and lifestyle modification are activities at community level that have a public health focus. Public health activities are the responsibility of both the health and social sectors. They should be provided by all health care professionals. But pharmacists can offer public health interventions more conveniently than other groups since they are easily accessible and recognized as experts in matters of health. Pharmacists are a trusted source for information and advice on health and medicines. They cannot, however, operate in isolation, but must accept joint responsibility with all health professionals to serve community and public health goals.

Supply chain services such as procurement, distribution and storage, as well as dispensing, are the basis of pharmaceutical care. For official recognition and reimbursement for interventions in the health care system, pharmacists usually need to comply with a wide range of rules relating to health care. Important aspects include terminology, standards, documentation, responsibility and accountability.

4.5.4   Level of involvement in the health care system

Three levels of decision-making relating to questions of health and therapeutic approaches can be differentiated. These form the framework in which the pharmacist contributes to patient care. Pharmacists can be involved at each level.

- · *At the patient level* the pharmacist needs to decide on issues of pharmaceutical care and triage, i.e. prioritization of care. The emerging task of pharmacists is to ensure that the patient's medicine therapy is appropriately indicated, the most effective available, the safest possible, is affordable and convenient. They also need to provide counselling on medicine taking and be involved in the monitoring of medicine usage.
- · *At the institutional level* of a health facility such as a hospital, clinic, managed care organization or pharmacy, management of therapy can include tools such as formularies, standard treatment guidelines and drug utilization reviews. Pharmacists are also important members of drugs and therapeutics committees.
- · *At the system level* planning, management, legislation, regulation and policy are the framework and basis on which any health care system will develop and operate. This applies equally to pharmacy related aspects of health care. The system level also includes standards of practice and mandates for pharmacy that are managed at national, federal, regional, state or district level depending on the country.

Pharmacists both now and in the future need to adapt their knowledge, skills and attitudes to these new roles. Areas of knowledge include pharmacology, pharmaceutics, pharmacoeconomics, pharmacoepidemiology, public health, medicine supply, medicine information management and rational medicine management. A focus on skills is imperative and includes the integration of basic science with clinical aspects of patient care, clinical skills, management and communication skills, active collaboration with medical teams and problem solving of medicine-related problems.

Attitudes displayed by pharmacists themselves have sometimes been a problem, hindering progress and recognition of the pharmacy profession by others. If they are to play a full and vital part in the management of pharmaceuticals pharmacists will need to adopt the essential attitudes required by health professionals working in this area: visibility, responsibility, accessibility in a practice aimed at the general population, commitment to confidentiality and patient orientation. Pharmacists will need both vision and a voice, and will need to integrate themselves fully into the health care team.

4.5.5 Training pharmacists in low income countries

Unfortunately, the importance of pharmacists in the health care sector is often under-estimated, particularly in low income countries. This may in part be due to the fact that pharmacy curricula have long been neglected, and have not always adapted to changes in health care systems. The role and function of pharmacists and pharmaceutical staff in low income countries needs to be reappraised, with a subsequent review of curricula adapted to the changing needs of the health sector.

Training and education at the pre-service level, as well as the strengthening of the academic structure, is essential if the sector is to develop to its full potential. Human resources development is a vital part of capacity building, and training and development in this area can make an important contribution to health care. Properly trained staff can promote rational medicine management, improve quality of care and lead to better use of resources.

WHO is assisting pharmacy schools in low income countries to develop pharmacy curricula that respond better to the challenges presented by health sector reforms and privatisation in the health sector. Some industrialized as well as low and middle income countries have started curriculum reviews, and have integrated the proposed changes into their undergraduate curricula. However, many curricula still lag behind the emerging needs of health care systems and have not adapted to new roles and functions.

4.6 Traditional healers

Estimates show that in many developing countries, particularly rural areas, up to 80 per cent of people visit and depend on traditional health practitioners and use traditional medicines (TM) as part of primary health care. Traditional healing has always been a component of health care. *Traditional medicine* plays an important role in the provision of health care in many developing countries: it has been fully integrated into the health systems of China, North and South Korea and Vietnam. Although allopathic medicine is available in most Asian countries, traditional medicine is still very popular, even in highly developed countries such as Japan. In Bhutan more than 2,990 medicinal plants are used in traditional medicine.

The use of traditional medicine, sometimes called *complementary and alternative medicine* (CAM), is also significant in many industrialized countries. Growing numbers of patients rely on CAM for preventive and palliative care. Traditional Chinese Medicine has been practised in Australia since the nineteenth century. In other industrialized countries the use of CAM is increasing: a survey found that 42 per cent of adults in the United States and 48 per cent in Australia have used it. In France, 75 per cent of the population has used complementary medicine at least once. The global market for traditional therapies is currently around US $ 60 billion per year and is increasing steadily.

TM usually involves local biological resources and knowledge about their actions and uses. Their widespread use and poor regulation has however created public health challenges regarding the safety, efficacy, quality and rational use of TM and CAM.

4.6.1 TM and CAM

TM includes diverse health practices, approaches, knowledge and beliefs incorporating plant, animal and/or mineral based medicines, spiritual therapies, manual techniques and exercises, applied singly or in combination to maintain well-being, as well as to treat, diagnose or prevent illness. TM is also often part of a wider belief system, and may be considered locally to be an integral part of everyday life and well-being.

WHO has defined TM as "the sum total of the knowledge, skills and practices based on the theories, beliefs and experiences indigenous to different cultures, whether explicable or not, used in the maintenance of health, as well as in the prevention, diagnosis, improvement or treatment of physical and mental illnesses".

In developing countries the term used most frequently is "traditional medicine". In industrialized countries, where the dominant health system is based on allopathic medicine, the term "complementary and alternative medicine" is usually used. CAM is increasingly used in parallel to or as a complementary system to allopathic medicine, particularly to treat and manage chronic and psychosomatic disease. Concerns about the compartmentalization and largely technical approach of Western allopathic medicine that ignore psychosocial circumstances due to financial and time constraints have led to a desire for more gentle, holistic, integrating and personalized health care.

Commonly used therapies of TM and CAM include Chinese medicine, Ayurveda, Unani, naturopathy, osteopathy, homeopathy, chiropractice (which may include herbal medicines), acupuncture and acupressure, manual therapies, spiritual practices and exercises. About 25 per cent of modern medicines have been derived from plants first used traditionally. Evidence is accumulating concerning the efficacy of certain TM therapies; for example, the efficacy of acupuncture in relieving pain and nausea is now well established.

4.6.2 Traditional healers

Traditional health practitioners are a valuable and sustainable resource that already exists in many developing world communities. The widespread use of TM can be attributed to easy accessibility and ready affordability. In Uganda, the ratio of TM practitioners to population is between 1:200 and 1:400; this contrasts with a ratio of allopathic practitioners to population of 1:20,000 or less. A large proportion of the population in many developing countries relies on traditional practitioners, including traditional birth attendants (TBA), herbalists, bone-setters and other healers. WHO estimates that TBA assist in up to 95 per cent of all rural births and 70 per cent of urban births in developing countries.

However numbers alone are not the whole story. Good quality of care in terms of outcome, both professionally defined and as perceived by patients, is essential. The training and utilization of traditional practitioners in primary health care, working in close collaboration with the formal health staff and infrastructure, can be an important contribution to effective, practical and culturally acceptable health care. Traditional healers are popular in many developing countries because they are embedded within wider belief systems. Since the TH workforce is an important resource for the delivery of accessible health care in developing countries, WHO has elaborated guidelines for the training of traditional health practitioners in primary health care.

A study conducted in South Africa explored the role of traditional healers: it showed that up to 70 per cent of patients would consult a TH as a first choice. There is increasing recognition that THs play an important role in preventing and controlling HIV/AIDS and other sexually transmitted infections (STIs). Most people in South Africa with STIs first seek help from traditional healers. Another study found that most traditional AIDS-educated healers now acknowledge that AIDS exists and take precautionary measures. However, many THs still believe that AIDS has evolved from older, mystical diseases that only TH can cure. In Africa, a number of HIV prevention programmes have involved THs whilst helping them to improve their skills in diagnosing, treating and counselling patients with HIV/AIDS and STI. An example of such a programme is illustrated in Box 4.7.

4.6.3 Traditional medicines

Medicinal plants are the oldest known health care product. In many countries their importance is still growing, depending on the ethnological, medical and historical background of the country. Medicinal plants also constitute an important source for

---

Box 4.7: Traditional healers and the management of sexually transmitted diseases in Nairobi, Kenya

"To describe the role of traditional healers in STD case management, in-depth interviews were held with 16 healers (seven witchdoctors, five herbalists and four spiritual healers) in four slum areas in Nairobi, Kenya. All healers believed that STDs are sexually transmitted and recognized the main symptoms. The STD-caseload varied largely, with a median of one patient per week. Witchdoctors and herbalists dispensed herbal medication for an average of seven days, whereas spiritual healers prayed. Thirteen healers gave advice on sexual abstinence during treatment, 11 on contact treatment, four on faithfulness and three on condom use. All healers asked patients to return for review and 13 reported referring patients whose conditions persist to public or private health care facilities. Thus, traditional healers in Nairobi play a modest but significant role in STD management. Their contribution to STD health education could be strengthened, especially regarding the promotion of condoms and faithfulness."

Source: Kusimba, J. et al. (2003) *International Journal of STD/AIDS* 14(3): 197–201.

---

pharmacological research and development. They serve as therapeutic agents or as bases and models for the synthesis of new medicines (see also chapter 13).

Definitions and regulations of herbal medicines vary considerably from country to country. Unfortunately, despite the increasing use of herbal preparations, the herbal market is often not adequately regulated and efficacy, safety and quality can often not be assumed. Whilst in some countries medicines derived from plants are well established and regulated, in others they are regarded as foods, and therapeutic claims are not allowed.

WHO has compiled a list of medicinal plants that are widely used in primary health care. For many traditional medicinal products formal, scientific, documented evidence of efficacy and safety is scarce. Yet these products have been "field-tested" for centuries, and much empirical knowledge has accumulated in communities and been passed on from one generation of healers to another.

Few countries have so far formulated policies on traditional medicine. This could result in the continuation of ineffective or unsafe practices that are harmful to patients. Unsustainable use of herbal medicines and plants can also lead to depletion of natural resources, sometimes called "the green pharmacy". The uncontrolled and unregulated commercial exploitation of natural plant resources is a threat to indigenous traditional knowledge.

4.7 Integrating traditional medicine and Western medicine

Doubts, suspicions and strong reservations about the role of traditional medicine still prevail amongst many health professionals and health administrators. Despite growing evidence of efficacy, cost-effectiveness and political commitment there is still resistance to providing support for traditional medicine, and to recognizing its benefits in some countries.

WHO supports the integration of traditional medicine into the mainstream health system, particularly for primary health care. As a result largely of public pressure, policy makers, health planners, administrators and health professionals have been challenged to show greater interest in traditional medicine. Policies and regulations are now needed to support and promote the rational use of traditional medicine.

A genuine interest in various traditional practices has, however, developed amongst many practitioners of modern medicine. At the same time growing numbers of practitioners of traditional, indigenous or alternative systems are beginning to accept and use some of the modern medical technologies. This will help foster teamwork amongst all categories of health workers within the framework of primary health care.

There are strong arguments in favour of including traditional healers fully into primary health care: the healers know the socio-cultural background of the people, and they are highly respected and experienced in their work. Economic considerations, the distances to be covered in some countries, the strength of traditional beliefs and the shortage of health professionals, particularly in rural areas, are also important factors supporting the greater integration of traditional healers.

4.7.1 Challenges of traditional, complementary and alternative medicine

Widespread and growing use of both TM and CAM has created significant public health challenges regarding policy, safety, efficacy, quality, access and rational use. The establishment of regulation and registration procedures has become a major concern in both developed and developing countries.

In 2000, only 25 countries reported having a national policy for traditional medicine, even though regulation or registration procedures for herbal products exist in nearly 70 countries. Many consumers use traditional medicine as self-care because there is a wide misconception that "natural" means "safe". As a result of unregulated availability they may be unaware of potential side-effects, and how and when herbal medicines can be taken safely. In 1996, for example, more than fifty people in Belgium suffered kidney failure after taking a herbal preparation which contained a toxic plant, *Aristolochia fangchi*.

Although traditional medicine has long been used, there is little systematic evidence regarding its safety and efficacy due to a context of cultural and historical conditions. Absence of evaluation has in turn slowed down development of regulation and legislation. There is also a lack of cooperation and sharing of information amongst countries as to the regulation of herbal products on the market.

The rational use of traditional medicine is hampered by lack of appropriate training for providers and of proper qualification and licensing schemes for THs. In addition, THs are not organized in networks of traditional practitioners, and their integration into modern primary health care systems is weak. The erosion of traditional lifestyles and cultures through external pressures, which may lead to loss of knowledge and traditional practices, adds further to concerns about the safety and rational use of traditional medicine. These concerns cast some doubt over its sustainable development.

4.8   Conclusion

This chapter has considered the differing roles of health professionals with regard to the management of pharmaceuticals in international health. It has emphasized the importance of effective teamwork if the patient is to obtain the maximum benefit from available medicines. We have suggested that the role of the pharmacist is crucial but under-utilized, and that pharmacists themselves still have much to do in order to change attitudes and develop services. The importance of education to this process is crucial.

We have considered the important role played by traditional healers with regard to medicines, and we have demonstrated the growing integration of traditional and Western medicine. TM is widely used, and is of rapidly growing health system and economic importance. To maximize the potential of TM, efforts need to be undertaken to formulate policies on TM, develop regulation and registration procedures, ensure quality and safety, study efficacy and promote rational use of traditional medicines.

Whilst traditional healers and herbal medicines represent an alternative for people in industrialized countries, they often are the only alternative for people in developing countries. In developing ountries, where more than 30 per cent of the population lacks access to essential medicines, the provision of safe and effective traditional medicines through qualified traditional healers could become a critical tool to increase access to health care. Nevertheless, there remain a number of concerns that cast doubt on its long term future.

Further reading

Anderson, S. (2002) "The state of the world's pharmacy: a portrait of the pharmacy profession", *Journal of Interprofessional Care* 16 (4): 391–404.

Cochrane Collaboration (2004) http://www.cochrane.org/index0.htm.

Correa-de-Araujo, R. (2001) "General principles of evidence-based pharmacotherapy". *The Consultant Pharmacist* Suppl. B: 3–5.

De Vries, T.P., Henning, R.H., Hogerzeil, H.V. and Fresle, D.A. (1994) *Guide to Good Prescribing*. WHO/DAP/95.1. Geneva: World Health Organization. http://www.med.rug.nl/pharma/who-cc/ggp/homepage.htm.

Green, E.C. (1999) "Involving healers", *AIDS Action*. Oct–Dec (46): 3.

Irwig, J., Irwig, L. and Sweet, M. (1999) *Smart Health Choices: How to make Informed Health Decisions*. St. Leonards NSW, Australia: Allen & Unwin.

Quick, J.D., Rankin, J.R., Laing, R.O., O'Connor, R.W., Hogerzeil, H.V., Dukes, M.N.G. and Garnett, A. (eds.) (1997) *Managing Drug Supply: The Selection, Procurement, Distribution and Use of Pharmaceuticals*. Second Edition. West Hartford, CT, USA: Kumarian Press.

Sackett, D.L., Straus, S.E., Richardson, W.S., Rosenberg, W. and Haynes, R.B. (2000) *Evidence-based Medicine: How to Practice and Teach EBM*. Second Edition. Edinburgh, New York: Churchill Livingstone.

SIGN 50: A Guideline Developer's Handbook (2001) Edinburgh: Scottish Intercollegiate Guidelines Network (SIGN). http://www.show.scot.nhs.uk/sign/guidelines/fulltext/50/index.html.

Timmermans, K. (2003) "Intellectual property rights and traditional medicine: policy dilemmas at the interface", *Soc Sci Med*. Aug 57 (4): 745–756.

WHO/EDM. *The role of the Pharmacist*. WHO/PHARM/)/599. Geneva: World Health Organization.

WHO/EDM. *Traditional Medicine Strategy 2002–2005*. WHO/EDM/TRM/2002.1. Geneva: World Health Organization.

Chapter 5
The Role of the Pharmaceutical Industry

Stuart Anderson

Box 5.1: Learning objectives for chapter 5

By the end of this chapter you should be able to:

· Describe the origins and development of the pharmaceutical industry.
· Explain the structure of the industry.
· List the principal partner organizations for the pharmaceutical industry.
· Describe the principal activities of the industry.
· Explain the role of the industry in medicine research and development.
· Explain the difference between primary and secondary manufacturing.
· List the key markets of the pharmaceutical industry.
· Explain the major implications of the globalisation of the pharmaceutical industry for developing countries.

5.1   Introduction

In this chapter we turn to the development of modern pharmaceuticals and to their production. As we saw in chapter one, for most of human history traditional medicines were the only forms of medicines available. The major developments in therapeutics came about only in the second half of the twentieth century. The main funders of research and development into new medicines and other health products have been governments, through public research laboratories and universities, and pharmaceutical manufacturers in industrialized countries. Recent decades have seen an enormous expansion of the pharmaceutical sector, and a consolidation into a relatively small number of very large transnational corporations (TNCs).

Today, the global pharmaceutical industry finds itself at the centre of debates about not only the availability and affordability of pharmaceuticals in developing countries but also the costs of medicines in developed countries. This chapter provides a basic introduction to the industry, and locates it within the broad context of international health. In doing so it explores six main themes: the origins and development of the pharmaceutical industry; the structure of the pharmaceutical industry; the pharmaceutical industry's partners; the principal activities of the industry;

the markets for the industry's products; and the role of the pharmaceutical industry in international health.

5.2   The origins and development of the pharmaceutical industry

As a manufacturing sector the pharmaceutical industry today is still very young. Its modern history spans only the last fifty years or so from the end of the Second World War. Yet pharmaceutical manufacturing has much earlier beginnings, and these origins have been different in different countries. In the United Kingdom, for example, a number of the businesses of apothecaries and chemists and druggists had, by 1870, already developed into manufacturing firms, typically for the large-scale production of chemicals and extracts of natural substances.

5.2.1   Foundations in the chemical industry

Mergers and acquisitions were taking place in the chemical industry long before the twentieth century, and today's pharmaceutical companies can often be traced back to these. For example, a flurry of activity followed the discovery of chlorine by Scheele in 1774. Numerous facilities for the manufacture of chlorinated lime were established, such as the Saint Rollox Chemical works in Glasgow founded by Charles Tennant in 1797. This facility became the United Alkali Company, which grew to become one of the four great companies which merged in 1926 to form Imperial Chemical Industries Limited (ICI). After the Second World War ICI established a pharmaceuticals division that split off in the 1980s as a separate company, Zeneca. In the early 1990s Zeneca merged with Astra, to form what became the fourth largest pharmaceutical company in the world, AstraZeneca.

There was another important strand in the development of the modern pharmaceutical industry, and that was the aniline dye industry. The industrial revolution had been powered by the carbonization of hard coal, which produced coal gas and coke but left coal tar behind. It did not take long to discover that coal tar was a rich source of dyes, and that these could be turned into almost any colour by relatively simple chemical modification. The first aniline dye was made by accident in 1856 by an English chemist, William Perkin, as he was trying to synthesise the anti-malarial agent quinine. Companies including Bayer, Hoechst, Ciba and Geigy all began business by manufacturing a range of products, including dyes and pharmaceuticals, from coal tar.

5.2.2   The therapeutic revolution

By the early twentieth century there were already in existence a substantial number of quite large pharmaceutical companies, with a diversity of origins in a variety of countries, mainly in Europe and America. Discoveries in the early decades of the century were to transform the industry. In 1910, the German scientist Ehrlich discovered that an organic arsenic-based substance called arsphenamine (later marketed as Salvarsan) was highly effective, despite its toxicity, in the treatment of syphilis. Over the following years a number of other effective chemotherapeutic agents appeared, including proguanil, chloroquine and the barbiturates.

But it was the discovery of the sulphonamides in the 1930s that triggered rapid expansion of the industry. The German company I.G. Farbenindustrie (IGF) patented Prontosil and many other azo-dyes containing a sulphonamide group. (IGF was dissolved by the allies after the second world war because of its involvement in the use of forced labour and concentration camps). Between them, pharmaceutical companies tested over 6,000 sulphonamides for therapeutic action. A large number of variants were marketed. The discovery of penicillin in 1928 provided another stimulus to the industry. Once the manufacturing process had been established during the course of the Second World War large numbers of pharmaceutical companies became involved in its manufacture, and in the search for further antibiotics.

After the end of the second world war a steady stream of new antibiotics appeared including erythromycin, the tetracyclines, the macrolides, and later the cephalosporins, the first of which was isolated in 1961. Amongst these were a number of medicines effective against tuberculosis: streptomycin, isoniazid and p-aminosalicylic acid (PAS). These were followed rapidly by the discovery of substances having a wide range of pharmacological activities, from psychoactive agents to anti-rheumatic therapies. Between 1952 and 1960, during the height of the "therapeutic revolution", around one thousand new pharmaceutical products were launched onto the British market alone.

5.2.3 Mergers and acquisitions

Although larger pharmaceutical companies have acquired smaller competitors from the beginning, and mergers have frequently occurred, the trend whereby very large pharmaceutical companies merge with each other to form vast TNCs was largely a feature of the 1990s. A few examples illustrate the change. The Swiss company Geigy was formed in Basel in 1758. It was over one hundred years before Ciba was created in the same city, in 1884, followed by Sandoz in 1886. By the middle of the twentieth century all three companies had successful product ranges. Ciba and Geigy merged as Ciba-Geigy in 1970, becoming Ciba again in 1992. Ciba then merged with Sandoz in 1996 to form Novartis, now the eighth largest pharmaceutical company.

Aventis is the result of a succession of mergers and takeovers over a period of many years. The major companies that have been absorbed into this corporation are summarised in Box 5.2. Hoechst started life as a manufacturer of dyes for the textile industry in the late 1800s: it began pharmaceutical production in 1883. It acquired Marion Merrel Dow and Roussel-Uclaf, themselves the products of successive mergers, to form Hoechst Marion Roussel in 1990. Rhone-Poulenc was

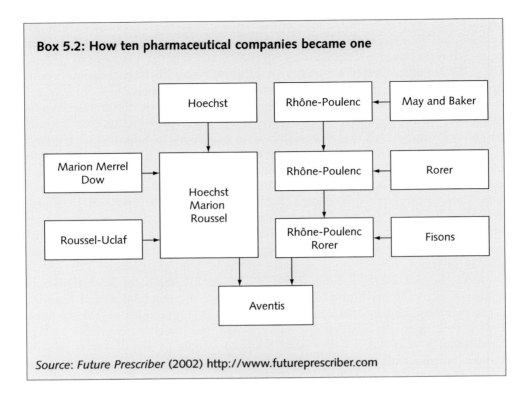

Box 5.2: How ten pharmaceutical companies became one

Source: *Future Prescriber* (2002) http://www.futureprescriber.com

founded in Lyon in France in 1895. An early acquisition was the British company May and Baker, and in 1995 it acquired Fisons. Finally, the merger of Hoechst Marion Roussel of Germany with Rhone Poulenc Rorer of France in 1999 produced Aventis, now the fifth largest pharmaceutical company.

Today, very few medium sized research-based pharmaceutical companies remain, particularly in industrialized countries. However, their place has largely been taken by biotechnology companies, which focus on a very limited range of pharmaceuticals produced by specific technologies. These have grown rapidly in recent years but remain small in comparison to the large TNCs. So the global pharmaceutical industry is characterised by a relatively small number of very big corporations, together with a large number of small companies filling particular niche markets. We now consider the structure of the industry in more detail.

5.3   Structure of the industry today

The structure and size of the pharmaceutical industry can be described in many different ways. Annual revenue is the usual indicator of both company and sector size, but other important elements of the structure are company size by market share, and the extent to which companies are reliant on a small number of high revenue

Box 5.3: Top fifteen pharmaceutical companies by value of sales

Sector rank 2001	Company	Recent acquisition/ merger	Country	Sector rank 1998
1	Pfizer	Warner Lambert	USA	5
2	GlaxoSmithKline	SmithKlineBeecham	UK/USA	12
3	Merck		USA	1
4	Astra/Zeneca		UK	4
5	Aventis		France/Germany	2
6	Bristol-Myers Squibb		USA	6
7	Johnson & Johnson		USA	9
8	Novartis	Ciba Geigy	Switzerland	8
9	Upjohn	Pharmacia	USA	–
10	Wyeth	American Home Products	USA	11
11	Eli Lilly		USA	8
12	Roche		Switzerland	10
13	Bayer		Germany	3
14	Schering-Plough		USA	14
15	Abbott		USA	13

Source: Adapted from Table 4.8. *The World Medicines Situation* (2004), Geneva: WHO.

products. But the pharmaceutical sector differs from other sectors in another important way, and that is its profitability.

5.3.1 Company size by revenue

If we measure size in terms of annual revenue, pharmaceuticals are by no means the largest industrial sector in the world. Even the largest pharmaceutical company in 1997 (the US firm Merck and Co) was only one hundredth in the league table of companies from all sectors. Its annual revenue was US $ 32,714 million. The largest company in the world, General Motors, had annual sales revenue nearly six times larger, at US $ 176,558 million. Even the twelfth largest company (Sumitomo in Japan) had revenue three times that of Merck.

Box 5.3 lists the top fifteen pharmaceutical companies with their 2001 and 1998 rankings based on value of sales, with recent mergers and acquisitions and their country of registration. The Box illustrates a number of features of the pharmaceutical sector: rank order has changed rapidly in recent years as a result of mergers and acquisitions; the fifteen companies are based in just five industrialized countries; and nine of the leading companies are American. Some fourteen pharmaceutical companies appear in the Fortune 500 list of the top 500 companies by revenue in the world.

Box 5.4: Relative profitability of pharmaceutical sector

Sector rank	Sector	Profit as % of sales (2000)
1	Pharmaceuticals	18.6
2	Banks	14.1
3	Communications networks and equipment	13.4
4	Semi-conductors and electronic components	12.7
5	Mines and petroleum extraction	10.8
6	Drinks	10.6
7	Telecommunications	10.6
8	Scientific, control and photographic equipment	10.4
9	Financial products	10.1
10	Computers and data services	9.7

Source: Fortune (2003) hppt://www.fortune.com

5.3.2   Relative profitability of sectors

But it is in their profitability that pharmaceutical companies stand out from other sectors. In the Fortune 2000 league table of profit as a percentage of sales by sector the pharmaceutical sector heads the list. The relative profitability of the pharmaceutical sector is illustrated in Box 5.4. On average, the profit as percentage of sales for pharmaceutical companies is 18.6 per cent. This is four and a half percentage points ahead of its nearest rival, banking, at 14.1 per cent. Typical profits as percentage of sales for other dynamic, thriving sectors such as telecommunications and petroleum extraction, are between ten and eleven per cent.

Whilst a significant proportion of sales revenue from the pharmaceutical sector is invested in research and development, large sums are devoted to the promotion of its products. But it is its very high level of profit as percentage of sales, coupled with the essential nature of those products that have set the pharmaceutical sector apart for particular criticism from many quarters. There is little doubt that if its profitability were more in line with other sectors much of this criticism would be averted.

5.3.3   Company size by market share

The pharmaceutical sector differs from other sectors in that market share is not dominated by a small number of big companies. Indeed, the company with the largest market share, GlaxoSmithKline, has only 7.1 per cent of the world market. The top twenty pharmaceutical companies by market share are shown in Box 5.5. The top five companies account for around 25 per cent of the market. Together, the top twenty companies account for just over 64 per cent of the world market.

Box 5.5: Top twenty pharmaceutical companies by market share

Rank	Company	Market share %	Rank	Company	Market share %
1	GlaxoSmithKline	7.1	11	Eli Lilly	3.0
2	PfizerWarnerLambert	6.9	12	AmericanHomeProducts	3.0
3	Merck and Co	4.9	13	Abbott Laboratories	2.6
4	AstraZeneca	4.6	14	ScheringPlough	2.4
5	BristolMyersSquibb	4.1	15	Bayer	1.8
6	Aventis	4.0	16	BoehringerIngelheim	1.4
7	Novartis	4.0	17	TakedaChemical	1.3
8	Johnson & Johnson	3.6	18	SanofiSynthelabo	1.3
9	Roche	3.2	19	Amgen	1.0
10	PharmaciaUpjohnMonsanto	3.1	20	ScheringAG	1.0

Source: Greener, M. (2001) *A Healthy Business: A Guide to the Global Pharmaceutical Industry*. London: Informa Publishing Group.

5.3.4 Role of blockbuster medicines

Most companies depend heavily on the revenue generated by a small number of medicines still enjoying the benefits of patent protection. In 1998 some thirty medicines reached global sales revenue of US $ 1.0 billion or more per year (the so-called "blockbuster" medicines). Usually there is a substantial time lag before value of sales reach this sort of figure. As a result a significant number of these are due to lose patent protection by the year 2005. Those with sales exceeding US $ 1 billion per year in 1999, and due to lose patent protection in 2005, are listed in Box 5.6.

Not surprisingly, the market is dominated by medicines for the treatment of diseases that are prevalent in developed countries, such as lipid lowering agents, medicines for ulcer healing and medicines for depression. It is clear that many of these conditions are potentially preventable. Yet the level of expenditure by the industry on research and development in these areas (estimated at almost US $ 13 billion in 2000) raises fundamental issues about society's overall investment portfolio and the incentives used to support it.

5.4 Pharmaceutical company partners

The global pharmaceutical industry is supported by a large number of organizations, including sub-contractors and partners, many of which are themselves very large global corporations. Most notable amongst these are the clinical research organizations (CROs) and the site management organizations (SMOs).

Box 5.6: Medicines with sales exceeding US $ 1 billion during 1999

Patent expiry	Brand name	Generic name	Main clinical use	Sales (US $ bn)
2001	Losec	Omeprazole	Ulcer healing	5.91
2005	Zocor	Simvastatin	Lipid regulation	4.49
2002/4	Claritin	Loratidine	Hay fever	2.70
2001/3	Prozac	Fluoxetine	Depression	2.61
2005	Lipostat	Pravastatin	Lipid regulation	1.80
2003/4	Ciproxin	Ciprofloxacin	Infections	1.69
2002/5	Zithromax	Azithromycin	Infections	1.30
2002	Klacid	Clarithromycin	Infection	1.25
2001	Zestril	Lisinopryl	Hypertension	1.22
2004	Diflucan	Fluconazole	Antifungal agent	1.00

Source: Adapted from Table 4.7. *The World Medicines Situation* (2004) Geneva: WHO

5.4.1   Clinical research organizations

CROs take on day-to-day responsibility for the management of clinical studies on behalf of their client companies. The number and size of CROs has increased enormously in recent years. By 1998 total sales to the industry had reached US $ 5.0 billion, and the global market increased by 31 per cent in 1999 alone. By 2000, some 30 per cent of pharmaceutical research and development expenditure was outsourced. Around 60 per cent occurs in the United States, with a further 36 per cent occurring in Europe. However, there is a limit to the extent to which pharmaceutical companies are prepared to outsource, with most believing that they need to retain a minimum level of in-house capacity.

The market leaders in the CRO market are currently Quintiles and Covanc. These had 24 per cent and 15 per cent, respectively, of the market in 1998. Other players are ClinTrials, Kendall, Parexel and PPD, which together account for a further 15 per cent of the CRO market. CROs typically provide a wide range of services, from pre-clinical services and clinical trials management, to data management and biostatistics and healthcare policy consulting. Nevertheless, they are not the only organizations providing support to the pharmaceutical industry.

5.4.2   Site management organizations

SMOs have expanded rapidly in recent years, and work for both pharmaceutical companies and CROs. They manage networks of hospitals, primary care practices and specialist clinics that collaborate to enrol patients and carry out clinical trials. They can provide ongoing on-site support, provide their sponsors with a single point of access to each site within the network, and can identify and assess potential investigators.

5.5 The activities of the industry

The industry today has many interests and activities, but its core business remains the development, production and marketing of new prescription medicines. In this section we examine the key activities of the industry.

5.5.1 Research and development

Pharmaceutical companies invest heavily in research and development (R&D). However, the purpose of that investment is to maintain profitability, and hence their position in the league tables of global and regional ranking. To achieve this they need to maintain a steady stream of therapeutic innovations in the drugs pipeline, and to sell them at high prices. The industry argues that the high cost of its products is due to the high cost of researching and developing new medicines. As a result R&D investment in pharmaceuticals is higher than in almost all other sectors. R&D investment by biotechnology companies is generally even higher, with some companies committing as much as 30 per cent of the total value of sales on R&D.

Box 5.7 illustrates the percentage of total sales committed to research and development by the top five biotechnology companies and the top ten pharmaceutical companies. The level of investment in R&D by biotechnology companies is greater than that of any pharmaceutical company. And although the actual spend on R&D between the largest pharmaceutical companies is comparable (between US $ 2.5 and 5.0 billion), the proportion of total sales this represents varies enormously, from five to seventeen per cent. However, at 17 per cent or less of total sales it is still a lot less than total expenditure for marketing and promotion.

There is no correlation between size of company based on sales, and proportion of sales spent on R&D. Indeed, some very large companies spend a surprisingly small proportion of sales revenue on R&D. Although Bayer spent US $ 1,017 million on R&D in 1999 this was just 3.5 per cent of sales revenue. Even the higher figures need to be put in perspective. Marketing and promotion generally account for between 30 and 35 per cent of sales, and the sector remains the most profitable sector overall.

It is frequently stated that the industry only invests in research on medicines where it estimates that there will be an income stream of at least US $ 300 million per year. The source of such a stream is inevitably seen as high income countries, with the result that most such medicines are for the treatment of diseases of affluence, such as obesity, depression and heart disease. In reality this target is actually achieved for only a minority of medicines, not least because all companies have effectively adopted the same strategy, and hence produce products that compete with each other.

Box 5.7: Pharmaceutical and biotechnology companies ranked by % of sales spent on research and development

Company	Region	R & D spent 2001 (US $ bn)	% of total sales
Biotech companies			
Chiron	US	0.34	30.0
Biogen	US	0.31	30.0
Genentech	US	0.52	22.0
Serono	US	0.31	21.2
Amgen	US	0.87	20.7
Pharmaceutical companies			
AstraZeneca	Europe	3.1	17.0
Pharmacia	US	2.6	16.5
Aventis	Europe	3.5	15.2
Pfizer	US	5.0	15.0
Roche	Europe	2.7	13.4
Novartis	Europe	2.9	13.3
GlaxoSmithKline	Europe	4.0	13.1
Bristol Myers Squibb	US	2.5	11.9
Johnson & Johnson	US	3.9	11.0
Merck	US	2.8	5.1

Source: Adapted from Figures 2.3 and 2.4. *The World Medicines Situation* (2004) Geneva: WHO.

Mechanisms for finding new drug entities have progressed enormously in recent years. Three strategies have come to dominate the development of new medicines. These are computer-aided drug design (CADD), combinational chemistry linked to high throughput screening (CC/HTS), and genomics. Yet there is little evidence so far that these strategies are producing the results anticipated. Two main reasons have been suggested as to why this might be. The first relates to management issues: large corporations are run by lawyers, accountants, salesmen and market strategists who often have little idea about the science involved. Large corporations often have difficulty in accommodating creative genius, and indeed organizational structures within many of them effectively prevent it.

The second is the profound complexity of the biology involved. The central problem is that most medicines interact with proteins, and the protein structures being investigated are very complex. There are simply no easy solutions still to be found. As one leading figure has said, what these strategies have done is to multiply the size of the haystack many fold without any increase in the number of needles. Whether the transnational pharmaceutical corporations that dominate the field will be able to deliver the new medicines needed by people in both the developing and developed worlds remains to be seen.

5.5.2   Pharmaceutical manufacturing

Although it tends to have a much lower profile than R&D or sales, manufacturing is the core activity of pharmaceutical companies. But it is generally very different to manufacturing in other sectors. It is the most regulated of industries, and pharmaceutical manufacturing is often several orders of magnitude more complex than that elsewhere.

It is important to clearly distinguish between primary and secondary manufacture of pharmaceuticals, a distinction that is of considerable importance to developing countries.

- *Primary manufacture* involves the processing of raw materials to make the active ingredient, usually a complex chemical entity.

This process is usually extremely complicated and very expensive, and usually involves the use of vast quantities of natural resources. For example, making just one kilogram of a recombinant biopharmaceutical may require the use of up to 30,000 litres of water. The antiviral drug enfuvirtide (Fuzeon) requires no fewer than 106 separate steps for its manufacture. The scaling up of such processes from the laboratory to the industrial scale represents an enormous technological and engineering challenge, and extensive pilot formulation and process development work are required. There are a limited number of manufacturers in the world for any given primary product.

- *Secondary manufacture* involves combining the active ingredient with a number of inactive ingredients (the excipients) to make a final pharmaceutical product.

This may take the form of any one of a wide variety of dosage forms, from tablets to inhalers. In most the active ingredient will be a tiny proportion of the total. Tablets will contain a range of excipients including diluents, lubricants to aid the manufacturing process, colours and flavours. Arrival at a final formulation will itself have been the result of extensive research and development work, and many medicine administration devices such as inhalers and injection pens are the result of extensive technological research in its own right.

5.5.3   Quality assurance

In developed countries there are few industries more regulated than the pharmaceutical industry. Such regulation is concerned not only with medicine safety issues before and after their introduction, but with the entire process of manufacture and distribution. Manufacturers must use high quality materials, and the manufacturing process has to adhere to exacting standards of Good Manufacturing Practice (GMP). These are detailed documents which form part of product licence applications. Basic principles for good pharmaceutical manufacturing practice are laid down by official bodies, and adherence to these is assured through regular inspection of manufacturing premises.

A quality assurance system has to be in place to ensure that every batch of a pharmaceutical is of the highest possible quality. This requirement means ensuring that all staff are properly trained, that premises and equipment are of the highest standard, and that documentation is comprehensive and meticulously completed. All this effort is designed to avoid any possibility of contamination or confusion between products, or any possible errors in labelling.

Not surprisingly, the investment in plant, premises and personnel required is great. As a result some companies contract out at least some stages of the process to third parties. This is particularly common for novel delivery devices involving the use of extremely expensive machinery. The same exacting standards apply to subcontractors as to the manufacturer.

The consequences of the exceptional nature of pharmaceutical manufacturing for developing countries are substantial. Because of the resources required, the complexity and the extent of the technical backup required, most but not all primary manufacturing is carried out in developed countries: India and China are also now major sources of such products. A number of developing countries are beginning to acquire the capacity for secondary manufacture, particularly for medium-technology dosage forms such as tablets, sometimes providing regional capacity to meet the demands of neighbouring countries.

5.6   Pharmaceutical markets

The market for the products of the global pharmaceutical industry is dominated by countries in the developed world. Box 5.8 illustrates the top ten pharmaceutical markets by sales revenue of 2000. The US accounts for around 50 per cent of the market, followed by Japan at over 18 per cent. But large middle income countries also represent a significant and increasing market for these products.

Collectively, developing countries account for only a tiny proportion of the business of the global pharmaceutical companies. But, as we have seen, this business is dominated by expensive medicines for diseases of prosperity: the pharmaceutical needs of developing countries are very different.

5.7   The pharmaceutical industry and international health

In reality the majority of the world's population does not have full access to the products of the global pharmaceutical industry. For many people, cultural and belief

Box 5.8: Top ten pharmaceutical markets

Rank	Country	2000 sales (US $ bn)	% of global sales
1	USA	149.5	52.9
2	Japan	51.5	18.2
3	France	16.7	5.9
4	Germany	16.2	5.7
5	UK	11.1	3.9
6	Italy	10.9	3.9
7	Spain	7.1	2.5
8	Canada	6.2	2.2
9	Brazil	5.2	1.8
10	Mexico	4.9	1.7
	Top Ten	279.3	98.7
	World Sales	282.5	100.0

Source: Adapted from Table 4.4. *The World Medicines Situation* (2004) Geneva: WHO.

factors mean that they place their faith in non-allopathic forms of medicine, such as Ayurvedic medicine in India, at least for non-life threatening conditions. Belief systems play an important role in defining the use of pharmaceuticals, as we have seen in chapter 3. But for many life-threatening conditions, such as AIDS, only allopathic medicine offers any real hope of successful treatment. However, for a great many conditions common in developing countries well-established medicines are readily available.

5.7.1   Generic medicines

Ninety five per cent of the medicines included in the WHO Model List of Essential Medicines (chapter 6) are no longer covered by patents. This means that, at least in theory, anyone can produce them and sell them under their own brand name or as generic medicines. It does not mean, however, that the medicines are either easy or cheap to manufacture. Nevertheless, they are ones for which the original manufacturers in developed countries are often willing to consider alternative models of supply to the export of finished products from those countries. A number of such models now exist.

Where the process of secondary manufacture is relatively straightforward, such as the mixing of ingredients and the tabletting of prepared granules, facilities are now available in many developing countries. These may be subsidiaries of transnational corporations, or locally owned firms. Established formulations and brand names are often also used under licence.

Primary manufacturing remains a distant goal for most developing countries, although some now have the capability of manufacturing simple therapeutic agents. And this capacity is increasing all the time. But for most countries pharmaceutical manufacturing means the preparation of proprietary medicines such as vitamin products and cough remedies, or the assembly of more complex medicines from imported raw materials.

Recent agreements have meant that some generic medicines can be manufactured whilst patent protection is still in force. Where pharmaceutical manufacturing is well-established, as in India, there are often problems associated with developing the markets for their medicines beyond their own borders. Current Indian law recognizes patents on how to make medicines but not on the medicines themselves. But generic manufacturers have been prevented from selling in other developing countries that have patent laws similar to those of developed countries.

5.8   Conclusion

The global pharmaceutical industry is an important and influential part of the market economy. Its purpose is to make profits for its shareholders, and at this it is extremely successful. These incentives encourage innovation and creativity, and again the pharmaceutical industry is very successful, spending considerable sums of money on R&D. As a market driven sector, it targets markets that offer the greatest potential for substantial returns, and this inevitably means focusing on diseases of developed countries.

As we saw in chapter 2, some diseases affect too few people for pharmaceutical companies to be prepared to invest large sums of money on research into them. There are over 5,000 such rare diseases, sometimes termed "orphan diseases", and collectively they affect millions of people in both the developed and the developing worlds. Finding treatments for these represents a colossal task. Even if it had the incentives to do so it is not one that the pharmaceutical industry can undertake alone. Extensive collaboration between the public and private sectors, between the state and industry, is required.

There is one key feature of the pharmaceutical industry that sets it apart from all other sectors: it does not operate in a true market environment. This situation is starkly illustrated in two imperatives: the need to make medicines available to people who could not normally afford them; and the need to develop medicines for unprofitable conditions, by means such as the orphan drugs programme. In addressing these issues it is of course vital that the pharmaceutical industry not only plays a key part but also considers that fulfilling its ethical obligations to humanity is in its own interests.

This feature of the industry is at the heart of concerns about how access to effective medicines by the people of developing countries can be improved, and we now explore these issues further in the chapters that follow.

Further reading

Abraham, J. (1995) *Science, Politics and the Pharmaceutical Industry*. London: UCL Press.

Davis, P. (1997) *Managing Medicines: Public Policy and Therapeutic Drugs*. Buckingham: Open University Press.

Greener, M. (2001) *A Healthy Business: A Guide to the Global Pharmaceutical Industry*. London: Informa Publishing Group.

Horrobin, D.F. (2000) "Innovation in the pharmaceutical industry". *Journal of the Royal Society of Medicine* 93: 341–345.

Kanji, N., Hardon, A., Harnmeijer, J.W., Mamdani, M. and Walt, G. (1992) *Drugs Policy in Developing Countries*. London: Zed Book.

Van Der Geest, S. and Reyndels Whyte, S. (1988) *The Context of Medicines in Developing Countries: An Introduction to Pharmaceutical Anthropology*. Dordrecht: Kluwer Academic Publishers.

Chapter 6
The Role of Governments

Rob Summers

Box 6.1: Learning objectives for chapter 6

By the end of this chapter you should be able to:

· Describe the aims, objectives and components of a national drug policy.
· List the main steps in the design and implementation of a national drug policy.
· List the main stakeholders, challenges and factors for success involved in the implementation of a national drug policy.
· Describe the drug supply management cycle.
· Describe the major constraints in effective procurement and ways to overcome them.
· List the main characteristics of an ideal drug supply management system.
· List the main in elements of drug legislation.
· List prerequisites for the functioning of a drug regulatory authority.
· List the elements required for effective enforcement of pharmaceutical laws and regulations.
· Describe the three levels of indicators used to monitor compliance with national drug policies.
· List the main indicator groups for each level.
· List additional factors which have an impact on the implementation of national drug policies.

6.1   Introduction

The governments of individual countries have a critical role to play in the proper management of pharmaceuticals. This chapter identifies and describes the main roles of governments in ensuring that their citizens have access to affordable quality medicines as and when they need them. It discusses four key roles of governments: the development of national drug policies; the development of health systems that support drug supply; the establishment of pharmaceutical legislation and regulation; and the creation of a research, monitoring and evaluation framework to support the process. We will consider each of these roles in turn.

Box 6.2: Components of a national drug policy

Legislation and regulation	Drug regulatory authority
	Quality assurance
	Standards of practice and education
Drug selection	Essential drugs concept
	Traditional medicines
Economic strategies for drugs	Affordable prices
	Sustainable financing
	Local production
Drug supply	Reliable supply systems
Rational use of drugs	Public and private sector
	Health professionals and patients
Supportive components	Research
	Monitoring and evaluation
	Human resources development
	Technical cooperation among countries

Source: Adapted from: Hogerzeil, H. and Laing, R. "National Drug Policy Implementation". Session 17, Drug Policy Issues course, Yogyakarta, Indonesia, 28th October–9th November 2001. http://dcc2.bumc.bu.edu/richardl/DPl2001/programme.htm.

6.2   The development of national drug policies

As we have seen, approximately one third of the world's population still lacks access to essential medicines. This is despite the large number of medicines available, and the fact that both medicine consumption and expenditure have increased dramatically over the past few decades. Factors that reduce access occur at various levels of the system. Safe and efficacious essential drugs (those drugs which meet the health care needs of the majority of a population) may not be available or affordable in poor countries at all, particularly to patients in the public sector.

Many factors contribute to this lack of access to essential drugs. Patient care points may be insufficient in number, or good quality drugs may fail to be supplied to them efficiently. Inappropriate prescribing and monitoring may also lead to a waste of resources and reduce the quality of care. The most common and successful way of identifying and correcting these factors is through a national drug policy.

6.2.1   The aim of a national drug policy

The main goal of a *national drug policy* (NDP) is to develop fully the potential that medicines have to improve health status within the available resources. The NDP comprises a commitment to goals, a guide for action and a common framework that

---

Box 6.3: Steps in the development of a national drug policy

Step 1:	Organize the policy process.
Step 2:	Identify the main problems.
Step 3:	Make a detailed situation analysis.
Step 4:	Set goals and objectives for the policy.
Step 5:	Draft the text of the policy.
Step 6:	Circulate and revise the draft policy-process is important.
Step 7:	Secure formal endorsement of the policy-legal, administrative.
Step 8:	Launch the national drug policy.

---

all stakeholders can follow. The NDP may cover a range of strategies implemented by the state and other role-players to reach the set objectives of the pharmaceutical sector. Commonly, these strategies have development, economic and health objectives.

An NDP should fall within the framework of the overall health care system, national health policy and health sector reform and should include both public and private sectors. The impetus to formulate an NDP may come from dramatic events such as a change in government that stresses needs for reform, or from a general need to improve the pharmaceutical situation.

6.2.2   The components of an NDP

An NDP generally aims to make safe and efficacious essential drugs available and affordable to the entire population, and to ensure that they are used appropriately by prescribers, dispensers and patients. It thus aims to provide equitable health care and to promote public health by making optimal use of the country's financial resources, and by strengthening the country's capacities in terms of management, education, information, legislation and infrastructure. The main components of an NDP are listed in Box 6.2.

6.2.3   Developing an NDP

The process of development of an NDP starts when political and professional leaders with vision and commitment are supported by internal and external stakeholders. Internal stakeholders from public and private sectors bring involvement and ownership. External stakeholders bring expertise, authority and financial support.

Development and implementation of NDPs need to be seen as part of a single continuous process. Indeed, implementation starts with preparation of the first draft of the policy. A mechanism for developing the policy, from organising the policy process up to the stage of its launch, has been described by WHO (2003, see reading list). There are eight steps in the development process. These are listed in Box 6.3.

Box 6.4: Steps in implementation of a national drug policy

Step 1:   Determine overall responsibilities.

Step 2:   Develop a master plan for NDP implementation and pharmaceutical sector development, including objectives and strategies, expected targets and outputs, and the responsible agency (e.g. ministry of health or health departments).

The lead agency may appoint a multidisciplinary NDP expert committee, with representatives of key role players, and a working group for each component of the policy. It will oversee and coordinate all components of the pharmaceutical system, and will monitor and assess implementation and achievement of targets and objectives. A unit with its own personnel and budget should be set up within the ministry/department for this purpose.

Step 3:   Identify other stakeholders in the policy-making process.

These will include drug regulatory agencies, district, provincial and regional offices, health facilities, other ministries (finance, trade, economic planning, education), private sector stakeholders such as industries and retailers, and third parties such as NGOs, consumer groups, academia and professional organizations.

Step 4:   Identify priorities and strategies for each component, and determine whether new enabling legislation will be needed.

Step 5:   Identify technical and financial resources.

6.2.4   Implementing an NDP

Once launched, the next stage is the implementation of the NDP. A number of the steps that need to be taken have been identified. Hogerzeil and Laing (reading list), for example, suggest that implementation should follow a plan of action that includes the five steps listed in Box 6.4.

6.2.5   Enabling factors for the implementation of an NDP

Key success factors for the implementation of an NDP include favourable political conditions, both inside and outside the country, a climate of economic and/or social reform, political commitment and skills at the highest levels, shared ownership among key groups, involvement of key health officials, support from the health care professions, adequate and scientifically sound data, technical expertise and capacity within the policy unit (ministry of health), clear and specific responsibilities, a technically sound policy and master plan, reasonable priorities, a realistic implementation timetable, and adequate financial resources.

6.2.6   Limiting factors for the implementation of an NDP

Obstacles to the implementation of an NDP can be of a political or technical nature. Experts should be consulted to ensure that the policy is technically sound. In terms of ethical challenges, it is important to avoid institutional or individual conflicts of interest, and to combat underlying corruption. The political landscape has to be taken into account. Continuous political analysis should take place at all stages of policy formulation and implementation. At the implementation stage, when political impacts become apparent, controllable and uncontrollable challenges may increase. Various groups will form to defend their own individual and group interests.

Good political strategies and tactics will therefore be required, including elements such as forming alliances and coalitions, coordination, public relations, bargaining and compromise, mobilisation of third party groups, and the creation of constituencies inside and outside government. It is important to recognise that technical and political role-players make decisions in different ways. Often these are at the very least blurred and may even be contradictory. Additionally, role-players from either group may change their decision-making paradigm, depending on the circumstances at the time.

It is much easier to design and provide the infrastructure for NDPs than it is to implement and maintain them successfully.

6.3   The drug supply process

Medicines differ from other consumer goods in a variety of important ways. Since they can save lives and improve health, and third parties may be responsible for payment, their use should be governed by moral and ethical considerations, to which any economic considerations should be secondary. On the other hand, it is difficult to assess the health benefits that drugs may achieve on an immediate, individual basis, as they cannot be compared with the potential benefits of other treatment options. Irrational use and irrational patient expectations are therefore common.

As drugs are generally costly, even small improvements in their management can produce large savings. For governments it is important that the drug supply process is efficient, whatever the health system in which it operates. Each stage of the drug supply process needs to be properly managed. The four steps of drug supply management are *selection*, *procurement*, *distribution* and *use*.

6.3.1   Selection

No country, health system or individual can afford access to all the drugs available on the market. Careful drug selection is crucial in order to achieve the best possible access to medicines for as many patients as possible. In the public sector, a list of *essential drugs* (those drugs that meet the health needs of the majority of the population) has been shown to be a good basis for the selection process. Since essential drugs have a major impact on the most common causes of morbidity and mortality, they are not only effective, but also cost-effective.

The selection of essential drugs is subject to market approval, reflects national morbidity patterns and takes into account criteria such as proven safety and efficacy, generic names and single compounds, rather than combination drugs, wherever possible. The WHO publishes a regularly updated model list of essential drugs (see chapter 7), which can be adapted by individual countries for their own needs.

Similarly, faced with an expanding pharmaceutical market and growing disease burden, cost-containment measures are increasingly being introduced into the private sector. Pharmaceutical benefit management schemes aim to optimise the use of available resources, for example by designing and adhering to their own formularies and basing their treatment guidelines on scientific evidence.

6.3.2 Procurement

Effective procurement ensures the availability of the correct drugs, in the correct quantities and dosage forms, at reasonable prices and at recognised standards of quality. Medicines can be procured through purchase, donation or manufacture. Governments need to ensure affordable prices, stable and adequate financing, appropriate quantification of requirements, a reliable supply system, and the quality, safety and efficacy of medicines, including the accuracy of product information.

Pricing is a vital component of drug procurement. The treatment of some infectious diseases such as malaria, tuberculosis and HIV/AIDS is characterised by resistance problems in the respective pathogenic agents. New effective medicines can be very costly, in comparison with tried and tested medicines. Hence, pharmacoeconomic analysis should be performed before new medicines are used and may then improve health care (this is described in chapter 12).

Governments can contain prices by removing or reducing taxes on essential medicines, promoting competition through generic or therapeutic substitution, taking advantage of TRIPS-compliant measures such as compulsory licensing and parallel importation (see chapter 7), and through price negotiations or controls. All countries which are members of the Organization for Economic Co-operation and Development (OECD) except the United States have some form of price control mechanism for pharmaceuticals.

Reliable financing and payment mechanisms are essential measures to procure medicines effectively. Regardless of the level of economic development, governments throughout the world tend to have limited efficiency in service delivery, which can hinder payment processes and affect procurement negatively. The problem is compounded by shortages and/or mal-distribution of funds used for pharmaceutical expenditure.

Mixed public/private financing options and collaboration with private companies are ways to address some of these problems (see section 6.3.5 on public-private collaboration below). Available funding should be prioritised to target the most important diseases and the poor and disadvantaged portions of the population. One

---

Box 6.5: Functions of a successful distribution system

· Maintains a constant supply of drugs.
· Keeps drugs in good condition.
· Minimises drug losses due to spoilage and expiry.
· Rationalises drug storage points.
· Uses available transport as efficiently as possible.
· Reduces theft and fraud.
· Provides information for forecasting drug needs.

---

way of achieving this objective is to "ring-fence" drug budgets, so that they cannot be used for other purposes.

Decentralised procurement can also help to address specific and/or local needs quickly. It may also increase accountability, as each centre is responsible for its own procurement practice. On the other hand, adherence to standard operating procedures, compliance with the essential drugs list, standardisation, equity and bargaining power through bulk purchases, will all tend to decrease, while administration procedures and the number of payment points will increase, through decentralised procurement.

6.3.3   Distribution

Within this step in the drug supply management cycle fall the activities required to receive drugs and move them safely, securely and in a timely manner to the points in the health care system where they are dispensed to patients. The basic functions of a successful distribution system have been described by Quick et al (reading list). They are listed in Box 6.5.

Various administrative options exist for managing distribution. It may be handled by government officials, whether they are part of the Department of Health or a different department charged with supplies and infrastructure, either at central or local level. In some instances it may be a contractor who handles procurement (see section on procurement above), or by the suppliers themselves, who deliver directly to dispensing points. In each case, safe storage, reliable records and accountability systems, efficient allocation of supplies and functional means of transport must exist at each distribution level. A typical example of a supply chain is illustrated in Box 6.6.

6.3.4   Use

Rational use of medicines means that patients receive medicines appropriate for their individual clinical needs, at correct dosages and for the correct period of time,

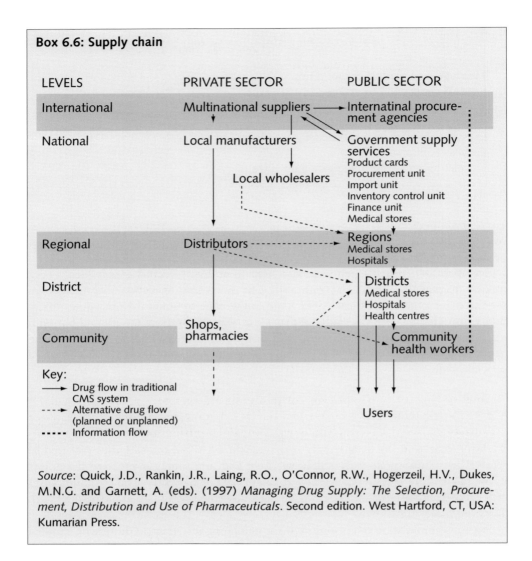

Box 6.6: Supply chain

| LEVELS | PRIVATE SECTOR | PUBLIC SECTOR |

International — Multinational suppliers — Internatinal procurement agencies

National — Local manufacturers — Government supply services
Product cards
Procurement unit
Import unit
Inventory control unit
Finance unit
Medical stores

Local wholesalers

Regional — Distributors — Regions
Medical stores
Hospitals

District — Districts
Medical stores
Hospitals
Health centres

Community — Shops, pharmacies — Community health workers

Key:
⟶ Drug flow in traditional CMS system
---⟶ Alternative drug flow (planned or unplanned)
····· Information flow

Users

Source: Quick, J.D., Rankin, J.R., Laing, R.O., O'Connor, R.W., Hogerzeil, H.V., Dukes, M.N.G. and Garnett, A. (eds) (1997) *Managing Drug Supply: The Selection, Procurement, Distribution and Use of Pharmaceuticals*. Second edition. West Hartford, CT, USA: Kumarian Press.

at the lowest cost to themselves and the community. The actual use of pharmaceuticals does not always comply with these requirements. It is influenced by factors such as drug availability, provider experience, economic interests, cultural factors and community beliefs. Widespread non-rational drug practices continue to affect health care outcomes and cause financial losses: prescribing of drugs which are not indicated, prescribing of too many drugs, use of wrong, ineffective, unsafe drugs, under-use of available drugs, and incorrect use of effective drugs.

Measures which can be taken by governments to promote rational drug use include the setting up of a multidisciplinary body to coordinate drug use policies, development of clinical guidelines, adoption of evidence-based treatment principles, undergraduate and in-service pharmacotherapy training of health professionals, elimination of perverse economic incentives to prescribers and dispensers, and regulation of marketing and promotion.

At national, regional and facility level, the formation of pharmacy and therapeutic committees can be encouraged to promote compliance with standard treatment guidelines and essential drugs lists and assist in monitoring and improving drug use patterns. We explore rational drug use more fully in chapter 10.

6.4   Health systems that support drug supply

The role of governments in health care can be seen from two different perspectives. According to the *social welfare or solidarity perspective*, the state should provide equitable health services to all as far as possible; whereas according to the *market economy or self-help perspective*, most aspects of health care should be left to the private market. Both systems have proved to have their drawbacks. Governments may lack the resources and efficiency to provide comprehensive health care, whilst private services may not be accessible to all groups of the economy.

6.4.1   Public-private collaboration

Various systems of public-private collaboration are therefore in use, particularly with regard to financing, procurement and distribution. In addition, not-for-profit organizations (such as NGOs) often make a major contribution to health care, particularly in remote areas of resource-poor countries. Examples of such partnerships are further discussed in chapters 2 and 13.

Public health care financing includes central, regional and local government budgets funded mainly by tax income. In addition, user charges (or, in emergency situations, development loans) can be considered as short term financing options. The promotion of medicine reimbursement as part of a compulsory public health insurance scheme would be a more sustainable option. Private financing includes out-of-pocket payments by individuals, private health insurance, community drug schemes, cooperatives, employers and financing through other non-governmental entities.

Huge savings can be made in many countries by increasing efficiency, reducing waste and combating crime and corruption. To a degree, public-private partnerships can contribute towards efficient procurement. Contracted companies may be more efficient than government in ensuring continuity of services, improving management information, and making timely payments to suppliers. On the other hand, continued cooperation with, and monitoring by, government are vital to ensure that the company's operations remain geared towards the objectives of the national drug policy and comply with strict specifications set by government.

Eight key factors for success in public/private collaborations have been described by Summers et al (reading list). These are listed in Box 6.7.

Box 6.7: Factors for success in public/private collaboration

· Contract objectives, terms, conditions and penalty and performance clauses must be clearly defined and understood.
· Deliverables, structures, systems and Standard Operating Procedures must be clearly defined.
· The relationship between the two sectors is that of a client and service provider, with the public sector the client and the private sector the service provider.
· The client must have the option to veto purchasing decisions by the provider, for good cause.
· The service provider must be selected on the basis of previous performance and capacity to carry out the required services.
· Collaboration should be considered only where appropriate human resources, infrastructure, an effective management information system and preparatory measures are available or can be readily provided.
· Funding must be guaranteed to give a sustained cash flow.
· Weak links and performance failure must be identified early.

Source: Summers, R.S., Möller, H., Meyer, D. and Botha, R. (1998) "'Contracting-out' drug procurement and distribution: experience with a primary distributor system in South Africa". *Essential Drugs Monitor* 5 and 6: 10–11. http://www.who.int/medicines/information/infmonitor.shtml

6.4.2   Cost analysis and budgeting in drug supply management

Costs are incurred at all stages of the drug supply management cycle. A number of specially designed management tools are available to analyse costs. The two most commonly used methods are VEN analysis and ABC analysis. These have already been considered in chapter 3.

Other methods that can be used include:

· *therapeutic category analysis*;
· *price comparisons* in various sectors and suppliers (useful resources include the MSH international drug price indicator guide published annually, and the WHO/HAI manual which provides a new approach to measurement of medicines prices at local or national level);
· *total variable cost analysis* aiming to model the cost impact of potential changes in the supply system;
· *lead time* and *payment time analysis*;
· *expiry date analysis* aiming to quantify the stock at risk of wastage; and
· *hidden cost analysis* in relation to the total value of purchased supplies.

Realistic pharmaceutical budgeting must be based on a rational selection and an

---

Box 6.8: Strategy for effective drug supply management

Structure: At each level, the following must be available:	Stock Financial resources Adequately trained personnel Equipment Community transport Buildings Reliable and timely deliveries of stock Good communication systems between participants Effective security measures
Process: The process will be achieved by	A standardised, practical policy Standard operating procedures Personnel evaluation In-service training Electronic communication Standardised documents

Outcome:

The outcome will be that adequate, reliable and safe medical supplies will be available to the community. The community will evaluate the personnel on the delivery of the service.

---

accurate quantification of drug requirements. Both aspects are vital in order to ensure cost-effective procurement of medicines, which is part of governments' responsibility to manage and use public funds so as to provide the best possible service. Selection should be based on pharmacoeconomic analysis in order to obtain maximum benefits for the amount spent on pharmaceuticals. Quantification should preferably be based on morbidity rather than past consumption, which may include drugs lost through theft, irrational use or inequitable allocations.

6.4.3   Strategies for effective drug supply management

It is possible to identify the main characteristics of an ideal drug supply management system. This helps us to focus on the key issues. One such "vision" for a drug supply management system five years into the future was developed at a workshop for primary level staff held by the Pharmacy Training and Development Project, School of Pharmacy, MEDUNSA.

A central drug store, whose function must be needs-driven, will supply the district hospitals, which will in turn supply clinics, mobiles, district surgeons, health facilities and other institutions. The central store will have ample stock, adequate personnel and pre-packing facilities. A strategy for effective drug supply management using the structure, process and outcomes approach is illustrated in Box 6.8.

An implementation plan is developed as part of this process, with the target of training at least two people from each primary care facility in drug supply management within a year.

6.5   Pharmaceutical legislation and regulation

One of the fundamental roles of governments is the enacting of legislation and the making of regulations across a very wide range of activities and issues affecting their people. Pharmaceuticals concern the entire population, affect the various stakeholders in different ways, and can involve important health risks. Informal control is insufficient to regulate the pharmaceutical sector. Laws and regulations are required to ensure the quality of medicines and of practice.

This legal framework must be enforced by an independent national drug regulatory authority. Areas to regulate include the manufacture, purchase, donation, import, export, distribution, supply, information, advertising and sale of drugs, and monitoring of adverse reactions. Multiple stakeholders with conflicting vested interests are involved in these processes. The importance of an international exchange of information between drug regulatory authorities has been stressed, not only to combat fraud and counterfeiting of drugs on an international scale, but also to address gaps in legislation, promote cooperation with the mass media for the benefit of public health, monitor the relationship between health professionals and the pharmaceutical industry and its effect on drug use, and promote active policies on rational drug use and consumer information.

It is important to note that any regulations drafted to control the use of medicines are only effective if they are enforceable.

6.5.1   Elements of drug legislation

We need to begin by defining the main elements of drug legislation. These have been described by Quick et al (see reading list) and are listed in Box 6.9.

6.5.2   Drug regulatory authority

A *drug regulatory authority* is a network that administers the full spectrum of drug regulatory activities. Its objectives are to establish a framework for drug regulation, and to operate a system of administration and enforcement so that drugs conform to acceptable quality, safety and efficacy (QSE) standards. The authority should carry out at least the following functions: Marketing authorization for new products and variation of existing authorization, quality control laboratory testing, monitoring of adverse drug reactions, provision of drug information, and enforcement operations.

Box 6.9: Elements of drug legislation

General provisions – definition of terms
Control of availability and marketing
· Drug registration (pre-marketing authorization)
· National Essential Drugs List/ National Formulary
· Scheduling, prescription, and dispensing authority
· Labelling
· Generic labelling, manufacturing and substitution
· Information and advertising
· Public education
· Imposition of fees
· Price control
· Special products (herbal medicines, orphan drugs)
Control of supply mechanisms
· Importation of drugs
· Exportation of drugs
· Controls, incentives, disincentives for local manufacture
· Control of distribution, supply storage, and sale
Drug control administration/drug regulatory authority
Powers to make rules and regulations
Repeals and transition provisions
Exemptions from provision of the law

Source: Adapted from Quick, J.D., Rankin, J.R., Laing, R.O., O'Connor, R.W., Hogerzeil, H.V., Dukes, M.N.G. and Garnett, A. (eds). (1997) *Managing Drug Supply: The Selection, Procurement, Distribution and Use of Pharmaceuticals*. Second edition. p. 94. West Hartford, CT, USA: Kumarian Press.

According to good regulatory practices, procedures and outcomes should be transparent to applicants, health professionals and the public, and the assessment process leading to pre-marketing approval of drugs should be of a reasonable duration without compromising quality, safety and efficacy.

Prerequisites for the work of a drug regulatory authority are the presence of political will and commitment, appropriate legislation, and adequate resources. These will include staff, premises, archiving, computers, and an expert advising body. There also needs to be an adequate fees structure that allows cost recovery. Accountability must be ensured at all levels of the process.

In the majority of countries the drug regulatory authority considers applications for drug registration on the basis of efficacy, safety and quality. A controversial "fourth hurdle" has been added to these three in some countries, in the form of a requirement for cost-effectiveness. In the majority of cases, cost-effectiveness is applied only for funding considerations, which is dealt with separately from registration for sale. In addition to general marketing approval, a special access scheme provides for the import or supply of unapproved therapeutic products for single patients under certain defined conditions or for the purposes of clinical trials.

Once a drug is approved and is on sale, ongoing *post-marketing surveillance* (PMS) mechanisms must be in place to monitor drug safety in the light of new scientific data and adverse reaction reports. Likewise, post-marketing quality assurance is important to ascertain that drugs on the market continue to meet QSE standards. Measures taken for this purpose include the standardization of the quality of raw materials and finished products, licensing of drug manufacturers, implementation of *GMP* criteria and inspections to monitor compliance with GMP, sampling and analysis of drugs, and inspection and supervision of drug distribution channels.

6.5.3 Advertising and promotion

Governments have an important role in regulating the promotion and advertising of medicines. Many countries now have regulations concerning these activities. The following example comes from Indonesia (see Slamet and Laing, reading list).

- Drug advertisements must be truthful, appropriate and must not mislead consumers.
- Advertisement of prescription drugs to the general public is prohibited.
- Prescription drugs may be advertised only in medical and pharmaceutical journals.
- Over-the-counter drugs can be advertised based on the information approved at the time of registration.
- Provincial drug inspectors monitor advertisements on a day-to-day basis.
- A committee evaluates advertisements before approval.
- Product samples may not be given to physicians.

For more information on advertising and promotion see chapter 5.

6.5.4 Enforcement

Pharmaceutical legislation and regulations must not only be laid down, but also enforced through administrative control or through court actions.

Laws and regulations must be functional both in their content and technically. The legislative framework must be kept updated to reflect both national policy and realities in the pharmaceutical sector. Difficulties in enforcement should be reported and evaluated to detect possible causes, related to technical defects in the law or to operational problems in its implementation.

Effective administrative enforcement requires a functional drug regulatory authority. Adequately trained personnel are needed for licensing, registration, monitoring, inspection and surveillance. In terms of physical infrastructure, office space, computers, software and office equipment, a quality control laboratory, and vehicles or distribution, inspection and enforcement activities are needed. The authority must receive adequate financing, which can be derived from tax income or from charges levied on manufacturers, importers and distributors at various stages of the licensing process.

In order to enforce drug regulations through court orders, the judiciary system must be functional: it must be independent and it must have the capacity to deal with court cases in a timely and competent manner.

6.6 Research, monitoring and evaluation

In order to detect possible weaknesses, the levels of compliance with the national drug policy must be monitored where possible. Indicators can be selected and used to measure changes, make comparisons and assess whether the targets are being met. If indicators are used they should be clear, useful, measurable, reliable and valid.

6.6.1 Categories of indicators

There are four main categories of drug policy indicators: background information, structural indicators, process indicators and outcome indicators. Background information provides data on the demographic, economic, health and pharmaceutical context in which drug policy is being implemented in a given country. Structural indicators provide qualitative information on whether the key structures, systems or mechanisms necessary to implement a pharmaceutical policy exist in a country. Process indicators assess the degree to which activities necessary to attain the objectives are carried out and their progress over time. Finally, outcome indicators measure the results achieved and the changes that can be attributed to the implementation of the national drug policy.

6.6.2 Indicators in practice

Some examples of indicators in each of the different categories are illustrated in Box 6.10. Detailed information on carrying out indicator studies can be found in two WHO documents: WHO/EDM/PAR/99.3 and WHO/DAP/93.1. Additional details are given in the list of further readings at the end of this chapter.

Based on the results of indicator surveys, possible underlying causes of functional deficiencies can be identified. Educational, managerial or regulatory interventions to improve pharmaceuticals management can then be designed and tested. On the national scale, the results will form the basis of strategic pharmaceutical policy planning.

Box 6.10: Categories and examples of indicators

BACKGROUND INDICATORS

Country information: population data (e.g. annual population growth, life expectancy) and economic data (e.g. GNP per capita, annual rate of inflation)

Health information: health status data (e.g. infant mortality rate) and health systems data (e.g. number of prescribers, total health expenditure)

Drug sector information: economic data (e.g. total pharmaceutical expenditure), human resources (e.g. number of pharmacists) drug sector organization (e.g. number of manufacturers, wholesalers, pharmacies) and number of drugs (e.g. registered medicines, essential medicines)

Key components:	STRUCTURAL INDICATORS (examples)	PROCESS INDICATORS (examples)
Legislation and regulation	Is there a checklist for carrying out inspections?	Number of drug outlets inspected, out of total number of drug outlets in the country.
Essential drug selection and registration	Is there a national essential drugs list?	Value of drugs from the national essential drugs list (EDL) procured in the public sector, out of total value of drugs procured in the same sector.
Drug allocation in the health budget/public sector financing policy	Has the public drug budget spent per capita increased in the last three years?	Value of public drug budget spent per capita in the last year, out of average value of the same budget during the past three years.
Public sector procurement procedures	Is there a system to monitor supplier performance?	Average lead time for a sample of orders in the last year, out of average lead time during the past three years.
Public sector distribution and logistics	Is the information recorded on the stock cards for a basket of drugs the same as the quantity of stock in store?	Average stockout duration for a basket of drugs in the central and/or regional stores in the last year, out of average stockout duration for the same basket in the past three years.
Pricing policy	Are drug prices regulated in the private sector? Is there a system for monitoring drug prices?	Average expenditure per prescription, out of average expenditure per prescription in the past three years.
Information and continuing education on drug use	Is there a national therapeutic guide with standardized treatments?	Number of prescribers having direct access to a (national) drug formulary, out of total number of prescribers surveyed.

> **Box 6.10: Categories and examples of indicators (continued)**
>
> ---
>
> OUTCOME INDICATORS
>
> *Availability of essential drugs*, e.g. number of drugs from a basket of drugs available in a sample of remote health facilities, out of total number of drugs in the same basket.
>
> *Affordability of essential drugs*, e.g. average retail price of standard treatment of pneumonia, out of the average retail price of a basket of food.
>
> *Quality of drugs*, e.g. number of drugs beyond the expiry date, out of the total number of drugs surveyed.
>
> *Rational use of drugs*, e.g. average number of drugs per prescription.
>
> ---
>
> *Source*: Brudon, P., Rainhorn, J.D. and Reich, M. (1999) *Indicators for Monitoring National Drug Policies*. WHO/EDM/PAR/99.3; Second Edition. Geneva: World Health Organization.
> http://www.who.int/medicines/library/par/indicators/who_edm_par_993.html.

6.7 Conclusion

This chapter has described some of the main functions that the governments of individual countries have in relation to the management of pharmaceuticals. As we have seen, these range from establishing health systems that support drug supply, to passing pharmaceutical legislation encompassing regulation, inspection and enforcement. We have emphasised that for low and middle income countries, and indeed for high income countries, the key to effective management of pharmaceuticals is the development, implementation and monitoring of national drug policies.

The short description of NDPs given in this chapter cannot do justice to the immense complexity of factors that affect their design, implementation and ongoing application. It is far easier to design and implement an NDP than it is to ensure its successful operation. This situation occurs because stakeholders soon identify ways of opposing and circumventing measures aimed at the common good, and subvert them to their own interests. They will carry out this process in many ways, going well beyond the health sector to do so.

In some instances they will wield their influence in economic issues, trade and industry to force drug policy changes. It is also quite common practice for internationally operating companies, in drug manufacture for example, to co-opt their own governments to pressurise less powerful or economically weaker countries to alter aspects of their NDP which may constrain profit-making and taking (see also chapter 8).

Further reading

Brudon, P., Rainhorn, J.D. and Reich, M. (1999) *Indicators for Monitoring National Drug Policies*. Second Edition. WHO/EDM/PAR/99.3. Geneva: World Health Organization.

Hogerzeil, H., and Laing, R. (2001) *National Drug Policy Implementation*. Session 17, Drug Policy Issues Course. 28th October–9th November 2001. Yogyakarta, Indonesia. http://dcc2.bumc.bu.edu/richardl/DPI2001/programme.html.

Quick, J.D., Rankin, J.R., Laing, R.O., O'Connor, R.W., Hogerzeil, H.V., Dukes, M.N.G. and Garnett, A. (eds). (1997) *Managing Drug Supply: The Selection, Procurement, Distribution and Use of Pharmaceuticals*. Second edition. West Hartford, CT, USA: Kumarian Press.

MSH (2002) *International Drug Price Indicator Guide*. Arlington, VA, USA: Management Sciences for Health.

Slamet, L. and Laing, R. (2001) *Legislation, Regulation and Quality Assurance Issues*. Session 9, Drug Policy Issues course, 28th October–9th November 2001. Yogyakarta, Indonesia. http://dcc2.bumc.bu.edu/richardl/DPI2001/programme.htm.

Summers, R.S., Möller, H., Meyer, D. and Botha, R. (1998) "'Contracting-out' drug procurement and distribution: Experience with a primary distributor system in South Africa". *Essential Drugs Monitor 5* and 6: 10–11. http://www.who.int/medicines/information/infmonitor.shtml.

WHO (1993) *How to Investigate Drug Use in Health Facilities. Selected Drug Use Indicators*. WHO/DAP/93.1. Geneva: World Health Organization.

WHO (2001) *How to develop and implement a national drug policy*. Second Edition. (updates and replaces *Guidelines for developing national drug policies 1988*) p. 27. Geneva: World Health Organization.

WHO (2003) "How to develop and implement a national drug policy". *WHO Policy Perspectives on Medicines* 6: 3.

WHO/EDM (2003) *The Rationale of Essential Medicines. Access, Quality and Rational Use of Medicines and Essential Drugs. Essential Drugs and Medicines Policy*. Geneva: World Health Organization. http://www.who.int/medicines/rationale.shtml.

WHO/HAI (2003) Medicines Prices: A New Approach to Measurement. WHO/EDM/PAR.2003.2. Geneva: World Health Organization and Health Action International. http://www.who.int/medicines/library/prices/medicineprices.pdf

Chapter 7

The Role of the European Union, National Assistance Agencies and NGOs

Reinhard Huss and Stuart Anderson

Box 7.1: Learning objectives for chapter 7

By the end of this chapter you should be able to:

· List the countries of the European Union.
· Describe the role of the European Union in the management of pharmaceuticals.
· List the main national aid agencies.
· Describe a sector wide approach.
· List four characteristics that define a non-governmental organization.
· Describe the principal aims and objectives of four non-governmental organizations.
· Outline the nature of global inequity in access to pharmaceuticals.
· Describe the purpose and scope of the Global Fund for AIDS, tuberculosis and malaria.
· Describe other initiatives that have been taken to improve global medicine supply systems.
· List sources of information for the supply of medicines in emergency situations.

7.1   Introduction

The globalization process, and its consequences for equity in the use of pharmaceuticals, is influenced by many stakeholders. As we have seen, these global actors can be divided into a number of categories: nation states with their national governments; the industrialised countries with their national aid agencies; the European Union; the pharmaceutical companies; the United Nations family of organizations as the highest global inter-governmental system; and the international NGOs as the representatives of global civil society.

In this chapter we discuss the role of some of the key players other than the industry (discussed in chapter 5) and the WHO (see chapter 8). We will discuss their potential role to promote more equitable pharmaceutical services, and stress the need for a political and ethical approach to be taken within the global context. The stakeholders considered here are the European Union; national aid agencies sponsored by some industrialised countries; and international NGOs.

Box 7.2: Countries of the European Union

Year joined	Country
1957	France, Italy, Germany, Belgium, The Netherlands, Luxembourg
1973	Denmark, Ireland, United Kingdom
1981	Greece
1986	Portugal, Spain
1995	Austria, Sweden, Finland
2004	Malta, Cyprus, Hungary, Poland, The Czech Republic, The Slovak Republic, Slovenia, Estonia, Latvia, Lithuania
(2007)	Bulgaria, Romania
applied	Turkey

Source: European Union (2004) http://www.europa.eu.int

7.2   The European Union

The *European Union* (EU) initially developed out of the Treaty of Paris in 1951 (which established the European Coal and Steel Community) and the Treaty of Rome in 1957 (which established the European Economic Community). Initially there were six members: France, Italy, Germany, Belgium, The Netherlands and Luxembourg. The EU then underwent four successive enlargements. Denmark, Ireland and the United Kingdom joined in 1973, Greece in 1981, Portugal and Spain in1986 and Austria, Sweden and Finland in 1995.

In 2004 the fifteen member states of the EU were joined by ten new members. These were Malta, Cyprus, Hungary, Poland, the Czech Republic, the Slovak Republic, Slovenia, Estonia, Latvia and Lithuania. These increase the land area covered by 34 per cent, and the population by 105 million, to 455 million. Bulgaria and Romania hope to join the EC by 2007. Turkey has applied for but is not currently negotiating membership. In order to join the EU, countries must fulfil the economic and political conditions known as the "Copenhagen Criteria". Under these, prospective members must be stable democracies, respect human rights, the rule of law and the protection of minorities; must have functioning market economies; and adopt the common rules, standards and policies that make up the body of EU law. The countries of the EU are listed in Box 7.2.

The EU is a unique blend of a supra-national system and an inter-governmental system. It has been subject to frequent tensions, because different governments often have very different aims. The *European Commission* (EC) is the executive branch of the EU. It wields considerable and increasing power because of its size and financial budget. Furthermore, the other institutions that control the administration in a state system (such as the parliament and the judiciary) are far less developed in the EU than is its administration.

7.2.1   The EU and the management of pharmaceuticals

The EC influences the management of pharmaceuticals in international health in four key areas. These are:

· Development assistance,
· Humanitarian assistance through the European Community Humanitarian Organization (ECHO),
· Operational research about health systems, and
· Regulation through the European Medicines Evaluation Agency (EMEA).

In the first three, this influence is mainly exercised through the financing of projects and programmes. The EC has taken an innovative approach to the regulation of medicines that has been watched with interest by the rest of the world. We will consider each of these in turn.

7.2.2   European development assistance

European development assistance is formulated and coordinated by the Directorate General for Development of the EC (known as DG Dev). DG Dev is responsible for maintaining cooperation between the EC and all developing countries, including sub-Saharan African, Caribbean and Pacific countries (ACP) and the Overseas Countries and Territories (OCT). The ACP states include 71 former colonies that, as a group, have a special cooperation agreement with the EU. This cooperation is funded through the European Development Fund. The OCT includes British, Danish and French overseas countries and territories.

The countries in the Mediterranean region, Asia, Latin America, Eastern Europe and the countries of the former Soviet Union are assisted through special cooperation and association agreements and programmes. This may also include areas of pharmaceutical management. In 2001 the EC set up a EuropeAid Co-operation Office. Its mission is to implement the external aid instruments of the EC which are funded by the European Community budget and the European Development Fund. Unfortunately the organization of European assistance is highly complex and lacks transparency.

7.2.3   European humanitarian assistance

ECHO is the humanitarian aid office of the European Commission. It funds not only the costs of humanitarian operations but also training and operational research. Sometimes activities are also co-financed together with other international funding agencies or the agencies of EU Member States.

Humanitarian action can be implemented at the request of international organizations and bodies, NGOs, member states, beneficiary third countries, or on the

Commission's own initiative. NGOs entrusted with implementing this action have to be either registered in the EU or in the beneficiary country. Humanitarian activities include the supply of medicines, and must be for the benefit of people in developing countries, ACP states and other third countries who are the victims of natural and man-made catastrophes.

The EC also funds community research with countries outside the EU. During the period between 2002 and 2006 the so-called sixth framework programme is being implemented. One part of this covers research for development, and includes the improvement of health systems from the central to the peripheral levels in ACP countries, Asia and Latin America.

7.2.4   European medicines regulation

The member states of the EU now share a common system for the evaluation of new medicinal products entering the European market. The *European Medicines Evaluation Agency* (EMEA) makes decisions which apply to the whole EU. These may have important implications for both the industry and for patients who stand to benefit from new therapies. EMEA is an important example of an initiative where national medicine regulatory authorities are partially replaced by a supra-national body. Although it has its headquarters in London, in many ways EMEA is a virtual agency: all EC national authorities are part of it, and all scientific evaluations are delegated to national assessors.

This initiative demonstrates some of the difficulties that arise when national agencies with different political, social, cultural and epidemiological contexts and history attempt to harmonize their regulatory frameworks. Some countries, such as Sweden, have very transparent approaches to regulation, whilst others, such as the United Kingdom and Germany, have less transparent ways of working. The process does not necessarily favour public health and citizens' participation, particularly as the number of decision-making bodies involved in the Europeanised process makes EMEA even less accessible to democratic participation than national agencies.

For any medicine regulatory body there is a balance to be struck between the need to make valuable new medicines available to the general population and the need to protect that same population against the dangers of exposure to the medicine. EMEA has been criticised for favouring a liberal agenda in which emphasis has been placed on facilitating the rapid marketing of new medicinal products in order to support the European pharmaceutical industry. There is far less emphasis on the development of a strong pharmacovigilance system to protect the European citizen from the potential risks and harm of new pharmaceuticals.

An additional danger is the increasing financial dependence of many national medicine regulatory authorities on fees levied on the pharmaceutical industry. This has also been the subject of some criticism. For example, the director of the German medicine regulatory authority (BfArM) has stated that: "It is German policy that medicine control is part of the policing laws, and as such you should not pay for the policeman who gives you approval. We come under the funds to protect public health".

7.3 The national aid agencies

7.3.1 The nature of national aid agencies

National aid agencies are the agencies of governments in industrialized countries that have the task of providing financial, personnel and technical services for low and middle income countries. Their constitution, nature and function vary considerably between countries. In Germany, for example, three separate government-controlled organizations provide these different types of assistance. Box 7.3 provides a list and short description of fourteen selected agencies.

The implementation of aid can be done in a number of different ways. Sometimes the agency itself implements it: sometimes it delegates it to an NGO, to one of the United Nations family of organizations, or to a private consulting agency. The support provided may be short-term humanitarian aid, or it may be medium-term development aid.

In the past, several agencies supported national pharmaceutical services either through pre-packed kits of medicines, which were regularly delivered to health facilities or donation of pharmaceuticals. Subsequently, a number of projects have been agreed between aid agencies and national governments to support specific elements in the system of supply, distribution and use of pharmaceuticals. These have included:

· training for health professionals;
· setting-up of national, regional and local supply and distribution systems;
· support for national medicines regulatory authorities; and
· assistance with operational research.

7.3.2 Sector-wide approaches (SWAps)

This previous project approach was not only inefficient in terms of resource use but also failed to recognise the importance of improving the functioning of the whole system, and of encouraging the recipient government to provide a coherent framework and policy for aid. As a consequence, assistance is increasingly being provided as health programmes, or as budgetary support for national governments.

Some aid agencies now demand that, as a condition for support, a transparent and reliable financial accounting and auditing system is installed by the recipient government, and that a coherent health sector programme exists. This is sometimes described as the *Sector-wide approach* or SWAp. This shift in approach has been led and promoted particularly by Dutch and British aid agencies.

7.3.3 The targeting of assistance

Another important development has been the increasing concentration of aid and assistance by governments on selected sectors and priority countries. A particular

Box 7.3: National aid agencies

Name	Country	Remarks
Australian Agency for International Development (AusAID)	Australia	AusAID is the Australian Government's overseas aid program. Australian funding is targeted at regional priority health needs, including maternal and child mortality, HIV/AIDS and infectious diseases such as tuberculosis and malaria, particularly in Papua New Guinea, East Timor, Indonesia and the Mekong countries. http://www.ausaid.gov.au/
Canadian International Development Agency (CIDA)	Canada	CIDA is an government owned development agency. Its role in the health sector is to promote health and nutrition including the reduction of infant, child, and maternal mortality and providing full access to safe water and sanitation services. Another focus is on risk groups most vulnerable to HIV/AIDS. http://www.acdi-cida.gc.ca/
Danish International Development Assistance (DANIDA)	Denmark	DANIDA is part of the Ministry of Foreign Affairs. Its actual role cannot be assessed from the available English websites. There is a general shift from project to programme and budget support. According to a health sector document from 1995 all aspects of comprehensive rational medicine management are supported by DANIDA. Since 2002 assistance is reoriented towards countries with long-term poverty reduction strategies concentrating on health and education and the role of the private sector. This includes emergency assistance and rehabilitation. http://www.um.dk/danida/
General Directorate for International Cooperation and Development (DGCID)	France	Development co-operation, particularly in health care, has long been a French commitment. Covering populations' essential needs and providing access for the greatest number of people to health care and education services are considered to be both engines of development and its end goal. France plays an active role in the Global Fund to fight AIDS, TB and Malaria (GFATM). http://www.france.diplomatie.fr/Thema/dossier.GB.asp?DOS=SOLIDARITYDEVEL
German Development Service (DED)	Germany	The German Development Service is mostly owned by the German government as non-profit making limited liability company, which works in the area of personnel cooperation. The DED supports request of partner organizations in the host country, which may include health and pharmaceutical services. http://www.ded.de/

German Technical Cooperation (GTZ)	Germany	The GTZ is a government owned corporation for international technical assistance. This includes programmes and projects to strengthen the health sector, especially quality improvement of and access to local health care, increasing health insurance coverage, improving access to essential medicines and emergency assistance. http://www.gtz.de/
German Development Bank (KfW)	Germany	The KfW finances on behalf of the German government social investments and related consultancies in the health sector. This includes the financing of essential medicines, particularly in the area of reproductive health, HIV/AIDS, tuberculosis and immunization, and the support for medical supply and distribution systems. http://www.kfw.de/
Directorate General for Development Cooperation (DGCD)	Italy	DGCD is part of the Italian Foreign Ministry. It plays an important role in fighting neglected diseases like leishmaniasis and African sleeping sickness with a major focus on Africa. http://www.esteri.it/eng/foreignpol/coop/index.htm
Japan International Cooperation Agency (JICA)	Japan	JICA is the official development agency linked to the ministry of foreign affairs. This includes technical and grant aid assistance. There is also a Japan Overseas Cooperation Volunteer (JOCV) programme and emergency assistance through the Japan Disaster Relief (JDR) team. The priorities in the health sector are the strengthening of district health services and the control of infectious diseases such HIV and tuberculosis. The Japanese government also provides official development assistance (ODA) as loans and contributions to international donor organizations. http://www.jica.go.jp/
Directorate-General for Development Cooperation (DGIS)	Netherlands	DGIS is part of the Ministry of Foreign Affairs. The Millennium Development Goals adopted by the United Nations are at the basis of the Dutch development policy. Within which DGIS devotes special attention to education, the environment, AIDS and reproductive health. DGIS believes that concentration and streamlining will boost the quality and effectiveness of development aid. Therefore the Dutch assistance prefers programme and budget support in these sectors. Moreover they concentrate just on a few poor nations, mainly in Africa. http://www.minbuza.nl/
Norwegian Agency for Development Cooperation (NORAD)	Norway	NORAD is part of the Norwegian Ministry of Foreign Affairs. Its aim is to improve the social, political and economic situation of the population in developing countries. NORAD has recently focused on HIV/AIDS, tuberculosis and vaccination but the overall aim is to strengthen the capacity of the health sector. http://www.norad.no/

Box 7.3: National aid agencies (continued)

Name	Country	Remarks
Swedish International Development Cooperation Agency (SIDA)	Sweden	SIDA is the implementing agency of the Swedish government and parliament. The priorities in the health sector are health system development, public health and the control of HIV/AIDS. SIDA supports the UN family of organizations and various recipient countries according to a jointly developed country strategy. http://www.sida.se/
Department for International Development (DFID)	UK	DFID is the UK Government Department to eliminate world poverty and promote sustainable development. The DFID has a strong commitment towards improving the health care of the most impoverished and marginalized. The UK has played a leading role in setting up the Global Fund to fight AIDS, TB and Malaria (GFATM). Otherwise, most of DFID's spending on HIV/AIDS goes directly to developing countries, as well as to support multilateral organization such as the United Nations. http://www.dfid.gov.uk/
United States Agency for International Development (USAID)	USA	USAID is the principal U.S. agency to extend assistance to countries recovering from disaster, trying to escape poverty, and engaging in democratic reforms. It is an independent federal government agency. Regarding the health sector, USAID is confronting global health through improving the quality, availability, and use of essential health services. USAID's strategy for global health seeks to stabilize world population and protect human health through programs in maternal and child health, HIV/AIDS, family planning and reproductive health, infectious diseases, environmental health, nutrition and other life-saving areas. http://www.usaid.gov/

aid agency may now have a presence in a low-income country, but not be working in the health sector. Priority countries change from time to time, but are usually listed on the websites of the aid agencies.

Emergency pharmaceutical assistance may be provided independent of such prioritization. However the necessary structure to link up with the country in need may be missing. Under these circumstances such aid is channelled either through the EU (in the case of a European donor), a national or international NGO, the UNHCR or the IFRC.

7.4   International and national NGOs

7.4.1   Definitions of NGOs

The term NGO is used in many different ways, and sometimes it is even used to include private for profit companies. We therefore need to have a clear definition of the term NGO before we can discuss their role in the management of pharmaceuticals. We can define an NGO using positive criteria. A non-governmental organization:

· is a formal organization;
· involves the voluntary participation of citizens;
· is independent of government decision-making;
· has a commitment to social welfare promotion.

Space limitations allow discussion only of a limited number of international and national NGOs. However, it will be evident from this definition that many more NGOs make an important contribution to the management of pharmaceuticals in international health.

7.4.2   The development of NGOs

NGOs can be further characterized by their stage of development and range of activities. Four stages of NGO development have been described by Korten:

Stage 1:   Local emergency relief and welfare provision,
Stage 2:   Community development projects,
Stage 3:   National development of policies and institutions,
Stage 4:   Global network for the advocacy of people's rights.

International NGOs such as MSF and OXFAM have evolved from stage 1 to stage 4, at which point these organizations are involved in activities from all four stages.

7.4.3 NGOs and the management of pharmaceuticals

NGOs are involved in a wide range of activities related to the management of medicines. These include:

· Emergency and routine health service provision
· Provision of support services
· Research activities
· Production activities
· Patients' interest representation
· Consumer representation
· Policy advocacy for public interests
· Public education

Support services include the supply and distribution of pharmaceuticals, management consultancy and the training of health professionals. NGOs are also active in research activities: an example is the operational research carried out by MSF in relation to the treatment of HIV patients. Research has been used by MSF as a strategy for overcoming the widespread opinion that effective treatment of such patients is not feasible in sub-Saharan Africa.

7.4.4 The origin of NGOs

The birth of an NGO is often linked to a market failure, in which the market failed to respond to an obvious need of a population. Many NGOs such as International Dispensary Association (IDA), Management Sciences for Health (MSH), OXFAM and MSF were created to fill needs arising in this way. The IDA, for example, was started in the Netherlands in 1972, because the global demand for a supply of essential, cheap and generic medicines was not met by commercial wholesalers. Today, the situation is very different, and we now find a variety of non-profit and profit-making wholesalers attempting to meet this demand.

New initiatives are occurring all the time. A recent example relates to the research, development and production of medicines for neglected diseases such as trypanosomiasis and leishmaniasis. The *"Drugs for Neglected Diseases initiative"* (DNDi) is another response of NGOs to an existing need which is not satisfied by the market system. NGOs move to take action wherever they judge it to be necessary.

Many so called failing or weak states cannot fulfil their role in the area of regulation and quality control of medicines. This may be due to lack of resources and finances but may also be the result of corruption and poor and ineffective organization of control institutions. In these cases NGOs may be the only option available to fill the gap. If a health care service wishes to provide cheap generic and good quality medicines in these circumstances, it may have to rely on the services and ethical standards of an NGO, such as IDA. In a recent example, a warning from the

NGO MSF revealed the circulation of counterfeit antiretroviral medicines in the Democratic Republic of Congo in February 2004.

7.4.5 NGOs and patient and citizen advocacy

International NGOs are also active representing patients' and citizens' interests. The initiative for neglected diseases referred to above is one example. There are many national NGOs who represent the interests of people suffering from specific diseases, health problems or the unwanted adverse effects of certain medicines. The American *AIDS Coalition To Unleash Power* (ACT UP) and the South African *Treatment Action Campaign* (TAC) have become widely known for their activities in the area of HIV/AIDS. They are also linked in an international network to advocate access to essential medicines for the treatment of HIV/AIDS and other diseases.

One of the problems associated with patients' interest groups is the possibility of "corporate capture". In this case a group is financed to a greater or lesser extent by one or more pharmaceutical companies in order to promote their pharmaceutical products. Patients' interest groups therefore need to be transparent about their funding, and to declare any possible conflict of interest.

Consumer associations try to represent the interests of citizens as pharmaceutical consumers. They therefore aim to promote transparency of information about medicines such as pricing, quality, labelling, safety, risks and benefits. There are many national NGOs dealing with consumer interests; they include the Consumers' Association in the United Kingdom. This organization publishes a Drugs and Therapeutics Bulletin for doctors and patients. The Drugs and Therapeutics Bulletin is a founder member of the International Society of Drug Bulletins. The latter's website provides an international list of drug bulletins, whose most important characteristic is their independence from the pharmaceutical industry. Unfortunately at present there are only three bulletins registered from the African continent.

7.5 Some non-governmental organizations

The main features of the main international non-governmental agencies and networks involved in the management of pharmaceuticals in international health are summarised in Box 7.4. In this section we provide additional information about some of them.

7.5.1 Health Action International

Health Action International (HAI) is a global network of NGOs that was founded in 1982. HAI was created as a result of estimates that up to 70 per cent of the medicines on the global market are either non-essential or undesirable products. It represents consumer and public interests. It focuses on campaigning for action on

Box 7.4: Important international NGOs and networks involved in pharmaceutical issues

Name	Description
International Federation of Red Cross and Red Crescent Societies (IFRC)	The IFRC is the world's largest humanitarian organization, providing assistance without discrimination as to nationality, race, religious beliefs, class or political opinions. Founded in 1919, the International Federation comprises 178 member Red Cross and Red Crescent societies, a Secretariat in Geneva and more than 60 delegations strategically located to support activities around the world. The IFRC carries out relief operations to assist victims of disasters, and combines this with development work. The focus are four core areas: promoting humanitarian values, disaster response, disaster preparedness, and health and community care including the provision of essential medicines. http://www.ifrc.org/
Global Treatment Access Campaign (GTAC)	The GTAC was founded at the International AIDS conference in Durban in 2000. It is a global network for communication and organization towards access to essential medications for HIV and other diseases. The core issues are exorbitant medicine prices, crippling debt, and a lack of sustainable public health strategies to meet the needs of those most affected by the AIDS epidemic, including women, children and the poor. GTAC is committed to working on these issues by creating and supporting partnerships between first- and third-world activist groups, and providing action tools and updates on current campaigns around the world. http://www.globaltreatmentaccess.org
Health Action International (HAI)	HAI is a global network of public interest groups involved in health and pharmaceutical issues. It was created because of estimates that up to 70% of the global medicines market are either inessential or undesirable products. HAI proposes that medicines should fulfil four criteria: to meet real medical need, to have therapeutic advantages, to be acceptably safe and to offer value for money. A core issue of HAI is the involvement of citizens in the regulation and control of public health care including the provision of essential pharmaceuticals. HAI is a response to the actual secrecy about health care regulation in many countries. The NGO promotes accountability and transparency of and democratic participation in the decision-making process of public administrations. It is assumed that such changes will improve the outcome of public health care interventions and make regulatory authorities more effective and efficient. A recent example for such efforts is the international medicine prices database developed by HAI and WHO. http://haiweb.org

International Dispensary Association Foundation (IDA)	IDA is probably the largest non-profit wholesaler of essential medicines and based in the Netherlands. It was founded in Amsterdam in 1972 and essential medicines to non-profit care initiatives in developing countries. As a supplier it simplifies the interface between many manufacturers in different countries and one customer and IDA assures the quality of the products. Therefore IDA is particularly useful for national and local distributors of pharmaceuticals and medical supplies who either do not have the time or the resources to deal with several individual manufacturers. IDA can also instantly provide the WHO Emergency Health Kits in case of disasters. http://www.ida.nl
International Network for the Rational Use of Drugs (INRUD)	INRUD was established in 1989 to design, test, and disseminate effective strategies to improve the way medicines are prescribed, dispensed, and used, with a particular emphasis on resource poor countries. The network now comprises 20 groups, 15 from Africa, Asia and Latin America, and other groups from the WHO/Department of Essential Drugs and Medicines Policy, the Harvard Medical School Department of Ambulatory Care, the Karolinska Institute in Sweden, the University of Newcastle in Australia, and a secretariat based in Management Sciences for Health in the United States. http://www.inrud.org/
Medecins sans Frontières (MSF)	MSF has started in France as an emergency medical aid organization in 1971 and still plays an important role in emergency assistance. Today it has branches in 19 countries and is active in over 80 countries to rebuild health services. As an NGO it wants to raise awareness for and address violations of human rights. It has been advocating that access to medicines has primacy in international agreements over commercial interests. MSF is one of the founder members of DNDi. http://www.msf.org/
Management Sciences for Health (MSH)	MSH was started in 1971 in the USA to help close the gap between knowledge and practice regarding health care problems in international health. The NGO works with policy makers, managers, providers and consumers. One of its four priorities is pharmaceutical management. It has become well known for its international medicine price indicator for essential medicines and the learning materials in the area of medicine management. http://www.msh.org/
Pharmaciens Sans Frontières (PSF)	PSF is the only humanitarian association in the world specialized in the pharmaceutical sector. It focuses on the improvement of health care for the most vulnerable populations. PSF's main goal is to assure the accessibility to quality pharmaceutical care for everybody, wherever, whenever, and as they are needed. The organization is active in the emergency and post-emergency/development sectors in 13 countries, and conducts pre-program assessments in 7 countries around the world. http://www.psci.org/indexuk.php3

Box 7.4: Important international NGOs and networks involved in pharmaceutical issues (continued)

Name	Description
OXFAM (originally Oxford Famine Relief) (OXFAM)	OXFAM started as a British NGO during the second world war to assist with famine relief in Europe. It is now an international group of 12 NGOs working in over 100 countries to fight against poverty and injustice. OXFAM implements many projects with local partners and wants people to know and claim their rights. It is also involved in advocacy to address global structural issues as a cause of poverty and injustice. Therefore it analyses international policies and practices which affect poor people. Among those are issues of drug patents, competition of generic medicines and pharmaceutical policy decisions taken by the EU or the WTO. http://www.oxfam.org/eng/

essential medicines by governments and UN agencies, and against industry products and practices that counter the concept.

According to HAI new medicines should fulfil at least four criteria. They should:

· meet real medical need;
· have therapeutic advantages;
· be acceptably safe; and
· offer value for money.

In order to achieve a better informed international consumer, HAI and WHO have developed a methodology to compare medicine prices between different countries. These results are published on the internet as an international medicine prices database.

One of the main issues on which HAI campaigns is the involvement of citizens in the regulation and control of public health care, including essential medicines. HAI can be understood as a response of international civil society to the secrecy and technocracy of health care regulation in many countries, especially in the European Union. The organization aims to strengthen accountability and transparency in the decision-making processes of public administrations, and to increase democratic participation in them. It is assumed that such changes will improve the outcome of public health care interventions and make regulatory authorities more effective and efficient.

7.5.2   International Network for Rational Use of Medicines

The International Network for Rational Use of Drugs (INRUD) was established in 1989 to design, test, and disseminate effective strategies to improve the way medicines are prescribed, dispensed, and used, with a particular emphasis on resource poor countries. It works through national multi-disciplinary groups from various sectors. It links the clinical and social sciences, and emphasises the behavioural aspects of medicine use.

INRUD promotes cooperation amongst donors and the sharing of experiences and technical expertise. It sponsors research projects to improve medicine use, organizes training, and publishes a newsletter twice yearly. This is accessible on the internet and distributed worldwide free of charge.

7.5.3   Management Sciences for Health

Management Sciences for Health (MSH) was founded in 1971. It works collaboratively with health care policymakers, managers, providers, and consumers to help close the gap between what is known about public health problems and what is done to solve them.

MSH seeks to increase the effectiveness, efficiency, and sustainability of health services by improving management systems, promoting access to services, and influencing public policy. It has recently been active in the countries of eastern Europe.

7.5.4   Healthy Skepticism

Healthy Skepticism (formerly known as the Medical Lobby for Appropriate Marketing, MaLAM) was founded in Australia in 1983. It is an international non-profit organization for health professionals and anyone else with an interest in improving health.

Its main aim is to improve health by reducing harm from misleading medicine promotion. Initially MaLAM concentrated on misleading promotion in developing countries, where the consequences may be worse than in industrialised countries because of a lack of regulatory controls and a lack of independent information. *Healthy Skepticism* has continued this work, and has expanded it to include inappropriate promotion from any country.

7.5.5   Other bodies

MSF and OXFAM are two of the other agencies actively involved in international policy advocacy matters. They are concerned with issues such as patent agreements and pricing issues, where international coordination is essential for effective action.

7.6   Equitable pharmaceutical supply and distribution

A number of initiatives have been introduced to address some of the key issues in the equitable supply and distribution of pharmaceuticals. These include the Global Fund for AIDS, TB and malaria, improving global supply systems, and supply in emergency situations. First, however, we review the nature of the inequity in pharmaceutical supply.

7.6.1   The nature of inequity

Global pharmaceutical services that are based on the values of justice and solidarity are not created by the market. As we have seen, leaving the supply of pharmaceuticals to market forces leads to a lack of access and inequity. Poor people have no financial means to buy the products that they need. The companies that produce and supply the products necessary have little interest in doing so, because they cannot make sufficient money under these circumstances. This is the important difference between a market and a democratic good which is made available to all members of society.

In addition, information is unequally shared between producers, suppliers, providers and consumers. Even well-educated consumers rarely know the truth about the quality of products provided. The occurrence of disease and disaster is unpredictable, often striking when people are least able to cope. The consequences for the people concerned are often a poverty trap, and individual measures of provision are simply not feasible for most citizens of the world.

7.6.2 The Global Fund for AIDS, tuberculosis and malaria

Several steps have now been taken to change this situation in low income countries. The global need for pharmaceuticals has to be translated into a quantifiable demand, which is then backed up by the financial means for meeting it. The development of a national health insurance scheme that includes all citizens and covers essential medicines can be one important step. However, many countries in the south may be economically too poor to take this step.

The creation of the *Global Fund for AIDS, tuberculosis and malaria* by the UN was a response to widespread criticism about the international apathy regarding the health crisis in most low income countries. Theoretically it offers the unique opportunity to finance pharmaceutical needs, at least in the area of some global health problems such as HIV/AIDS, malaria and tuberculosis. The fund aims to give priority to countries and regions with the greatest needs, measured in terms of burden of disease, and with the least ability to meet these needs, measured in terms of available financial means. Proposals to the fund are submitted through a Country Coordinating Mechanism (CCM) which is made up of representatives from government, bilateral and multilateral donors and non-governmental organizations which are active in the country.

One major problem of the global fund is the shortfall between proposals submitted and worthy of approval and the actual commitment of the international community to provide sufficient financial means for the fund. OXFAM has criticised this situation as a typical example of so-called donor fatigue. It reported that "in 2001, the total GDP of the 30 OECD countries amounted to approximately US $ 25 trillion. If these countries had met their international obligations in allocating 0.7 per cent of GDP in aid to developing countries, this would have generated approximately US $ 171 billion. Although the UNAIDS figure of US $ 10 billion for combating AIDS seems like a large sum, it is equivalent to four days of global military spending, ten days of OECD agricultural subsidy, or the cost of 100 Eurofighters".

Another important problem is the state of governance and existing health systems in countries with the most urgent needs. For example, the government of the Democratic Republic of Congo currently provides almost no financial funds to its health care system, although the country is very rich in mineral resources and continues to export illegally large quantities of these resources without any benefit to its population. Unfortunately global fund programmes may not achieve their objectives under such circumstances, if the problems of governance and funding of health care systems are not addressed properly.

The global fund should nevertheless be seen as an enormous opportunity to strengthen health care systems in developing countries. It may even be a first step towards a much greater vision. However, whether such a development can be taken a giant leap further and become a global back-up insurance for essential health services including medicines, remains to be seen.

7.6.3 Improving global supply systems

Another step that can be taken is the improvement of the global supply system. Several organizations have been created since the foundation of IDA as one of the first international suppliers functioning on a non-profit basis. Others can be identified on the international price indicator website of MSH and WHO. Some of these organizations also provide pre-packed standardized kits available for immediate supply in case of emergencies.

National, regional and local distribution systems have been developed in recent years. In the past many low-income countries had state-run distribution systems: many of these failed, mainly because of lack of funds and corruption. In recent years United Nations organizations such as the World Bank and the WHO, the EU with its aid programme, several National Aid Agencies and international NGOs have assisted with the development of delivery systems for pharmaceuticals and medical supplies. These can either be for-profit or not-for-profit organizations, and need to be regulated and controlled by the state so that they serve the public interest.

In several countries such as the Central African Republic, Cameroon or Rwanda public organizations have been created to provide and distribute essential medicines to public health facilities. These organizations have an autonomous status so that management and financing of medicines can be independent of other government interests. It remains to be seen whether they will be successful under circumstances where the state is too weak or is unable for other reasons to control and reinforce regulations for pharmaceutical services.

Church-related NGOs are often active in international or national medicine supply. An international example is Action MEDEOR in Germany. National and regional examples exist in many African countries such as the Central African Republic, Cameroon, Democratic Republic of Congo, Kenya, Tanzania and Uganda. Some are partnerships between different churches and they may collaborate with local NGOs. They were often created as a response to difficulties in obtaining good quality and low price generic medicines in these countries. Some of these supply organizations are financially dependent on foreign funders, whilst others have become financially self-sustainable and even provide services to the public and private sectors.

7.6.4 Supply in emergency situations

Market forces do not ensure the supply and distribution of pharmaceuticals in emergency situations. This is a particular case of market failure. Disaster preparedness therefore needs to be organized at all levels: this includes the local, regional and national levels. Emergency pharmaceuticals need to be stored at each level, according to past experience with particular types and frequencies of disasters.

Important documents to assist with disaster preparedness are the *New Emergency Health Kit* of WHO and the *CD Health Library for Disasters* published by WHO and PAHO. Useful information is also available on the website of the *Sphere Project*. This project was started in 1997 to develop a set of universal minimum standards in core areas of humanitarian assistance.

At the international level many national aid agencies provide such assistance. These include the EU with its ECHO programme, the UNHCR, and several international NGOs. This may include immediate emergency relief and rehabilitation programmes.

7.7   Conclusion

This chapter has considered the roles of some of the many other organizations besides national governments, WHO and the pharmaceutical industry that play a crucial role in the management of pharmaceuticals in international health. Collectively they make an enormous contribution to ensuring access to pharmaceuticals to many more people than would otherwise be the case.

The EU plays a particular role because of its unique blend of supra-national and inter-governmental systems. In particular, the *EMEA* is an important example of an initiative where national medicine regulatory authorities are partially replaced by a supra-national body and virtual agency. This initiative has important implications for the future in the light of the wider global problems in medicine regulation. How potential conflicts of interest between European citizens and the pharmaceutical industry are dealt with will be closely scrutinized by other countries.

Some national and international NGOs such as OXFAM and MSF are important advocates of people's rights in the globalization process. This advocacy needs to make sure that the changes sought are not limited to the economic sector, but contribute to the improvement of social conditions, and are based on justice, solidarity and ethical principles.

But the existence of so many diverse organizations, with very different aims and objectives, accountabilities and sources of funding, also creates many challenges. This chapter has demonstrated the complexity of the civil society in which pharmaceuticals are managed, and highlights the importance of co-ordination and co-operation between them. We return to this issue in a later chapter.

Further reading

Abraham, J. and Lewis, G. (2000) *Regulating Medicines in Europe*. London: Routledge.

Chang, H.-J. (2002) *Kicking Away the Ladder: Development Strategy in Historical Perspective*. London: Anthem Press.

Chetley, A. (1995) *Problem Drugs*. Health Action International. London: Zed Books.

Green, A. and Matthias, A. (1997) *Non-governmental Organizations and Health in Developing Countries*. London: MacMillan Press.

Huss, R. (1995) "Pharmaceutical consumer co-operatives: The third path?" *World Hospitals and Health Services* 31: 13–15.

Kawasaki, E. and Patten, J.P. (2002) *Drug Supply Systems of Missionary Organizations. Identifying Factors Affecting Expansion and Efficiency: Case Studies from Uganda and Kenya*. Report prepared for WHO/EDM. WHO#HQ/01191467. Geneva: World Health Organization.

Korten, D. (1990) *Getting into the Twenty First Century*. West Hartford: Kumarian Press.

OECD Health Data 2002 (2003) Paris: Organization for Economic Cooperation and Development.

OXFAM (2002) *False Hope or New Start? The Global Fund to Fight HIV/AIDS, TB and Malaria*. Oxford: OXFAM.

Vogel, D. (1998) "The globalization of pharmaceutical regulation". *Governance* 11: 1–22.

Walt, G., Pavignani, E., Gilson, L. and Buse, K. (1999) "Managing external resources in the health sector: Are there lessons for SWAps?" *Health Policy and Planning* 14: 273–284.

WHO (1999) "Globalization and access to drugs". *Health Economics and Drugs*. DAP Series No. 7. WHO/DAP/98.9 Revised. Geneva: World Health Organization.

WHO (1998) *The New Emergency Health Kit 98*. Second edition. WHO/DAP/98.10. Geneva: World Health Organization.

Chapter 8
The Role of International Organizations

Rob Summers

Box 8.1: Learning objectives for chapter 8

By the end of this chapter you should be able to:

· Describe the role of international organizations in the management of pharmaceuticals.
· List organizations affiliated with WHO or the United Nations that play a role in international drug management.
· List industry organizations that play a role in international drug management.
· Define the term "essential drug".
· List WHO selection criteria for essential drugs.
· Describe main elements of the review and revision of the WHO-EDL.
· Explain the relationship between treatment guidelines and the EDL.
· Describe recent developments in drug pricing to improve equity and access.
· Define and list current controversies in differential pricing.
· List international trade agreements that affect access to medicines.
· Discuss the implications of TRIPS for the supply of medicines.
· List provisions in TRIPS that act as safeguards for individual country interests.

8.1 Introduction

International organizations have multiple functions in drug management, including the promotion of concepts which aim to rationalise drug supply and use, the provision of finance, the promotion of consumer and public health interests, the provision of drug information and of technical assistance in specialised fields.

This chapter identifies the key international organizations involved; they include not only WHO and the UN but private foundations and organizations representing the pharmaceutical industry. It describes the concept of essential drugs and it considers the process of drug procurement, including the key issues of pricing, marketing and promotion. And it reviews the various international agreements relating to pharmaceuticals, including the agreement on trade related aspects of intellectual property rights (TRIPS). We will consider each of these issues in turn.

8.2   WHO and other UN organizations

The WHO, based in Geneva, has several divisions that play a role in improving international pharmaceuticals management. The WHO headquarters structure is illustrated in Box 8.2.

8.2.1   The Essential Drugs and Medicines Policy Division

A Drug Policies and Management unit (DPM) was first established in the WHO in 1977 in order to implement a new approach to drugs policy. In 1978 the World Health Assembly agreed to launch an action programme on essential drugs, and in 1981 the DPM was replaced by the Action Programme for Essential Drugs, known as DAP (Drug Action Programme). Since then it has undergone a number of further changes in name and function, but throughout it has played a leading role in promoting the essential drugs concept.

Today the Essential Drugs and Medicines Policy division (EDM) provides the impetus and secretarial support for the implementation of the essential drugs programme. It:

· Revises the Model List of Essential Drugs on a regular basis.
· Publishes documents on practices and methods for investigating and improving drug use.
· Publishes the *Essential Drugs Monitor* newsletter on current developments around the world.
· Convenes expert committees, holds workshops and contributes to short courses worldwide and supports country-specific programmes.
· Is responsible for the quality certification scheme and good manufacturing practices.
· Promotes quality assurance and safety of medicines by supporting regulatory processes and providing norms and standards.
· Produces the International Pharmacopoeia.

Other WHO offices which deal with specific disease areas, as well as the regional offices and individual country programmes, are also involved in international pharmaceutical management.

8.2.2   Other UN organizations

Other UN organizations involved in health care include:

· The United Nations Conference on Trade and Development (UNCTAD);
· The United Nations Drug Control Programme (UNDCP);
· The United Nations Development Programme (UNDP);
· The United Nations Population Fund (UNFPA);

Box 8.2: Structure of WHO headquarters

Director-General
Director
Advisers
Ombudsmen
Legal Counsel
Internal Audit and Oversight
Media and Communications

Representatives
of the Director-General
Polio Eradication
Health Action in Crises

Link to Regional Offices

ADG
HIV/AIDS, TB and Malaria

Director
HIV/AIDS
Director
Stop TB/
Partnership Secretariat
Director
Roll Back Malaria/
Partnership Secretariat
Director
Strategic Information

ADG
Communicable Diseases

Director
CD Surveillance and Response
Director
CD Control, Prevention and Eradication
Director
Special Programme for
Research and Training in
Tropical Diseases

ADG
Noncommunicable Diseases
and Mental Health

Director
Noncommunicable Disease
Prevention and Health Promotion
Director
Management of
Noncommunicable Diseases
Director
Injuries and Violence Prevention
Director
Nutrition for Health and Development
Director
Mental Health and Substance Dependence
Director
Tobacco Free Initiative
Director
Surveillance

ADG
Sustainable Development
and Healthy Environments

Director
Protection of the
Human Environment
Director
Food Safety
Director
MDGs, Health and Development Policy
Director
Ethics, Trade, Human Rights and Law
Director
Country Focus

ADG
Health Technology and
Pharmaceuticals

Director
Essential Drugs and
Medicines Policy
Director
Essential Health
Technologies

ADG
Family and Community Health

Director
Child and Adolescent Health
and Development
Director
Reproductive Health
and Research
Director
Gender and Women's Health
Director
Immunization, Vaccines
and Biologicals

ADG
Evidence and Information
for Policy

Director
Evidence for Health Policy
Director
Health Financing and Stewardship
Director
Health Service Provision
Director
Research Policy and Cooperation
Director
Health Information
Management and Dissemination

ADG
External Relations
and Governing Bodies

Director
Governance
Director
Government, Civil Society and
Private Sector Relations
WHO Offices at the:
United Nations, New York
African Union, Addis Ababa
European Union, Brussels
World Bank, Washington DC

ADG
General Management

Director
Programme Planning,
Monitoring and Evaluation
Director
Human Resources Services
Director
Office of the Comptroller
Director
Support Services, Procurement and Travel
Director
Information Technology and
Telecommunications
Director
Global Management System
Director
Security Coordination

World Health Organization

5 November 2003
Implementation of this structure is being phased in

· The United Nations Children's Fund (UNICEF);
· The Joint United Nations Programme on HIV/AIDS (UNAIDS);
· The United Nations Industrial Development Organization (UNIDO); and
· The World Bank Group.

The World Bank Group consists of five closely associated institutions owned by member countries that carry ultimate decision-making power. The term "World Bank" refers specifically to two of the five, the *International Bank for Reconstruction and Development* (IBRD) and the *International Development Association* (IDA).

The *International Monetary Fund* (IMF) is an international organization of 184 member countries. It was established to promote international monetary cooperation, monetary exchange stability and orderly exchange arrangements; to foster economic growth and high levels of employment; and to provide temporary financial assistance to countries to help ease balance of payments adjustments. Today, the IMF's operations have adapted to an evolving world economy and include surveillance, financial assistance, and technical assistance.

International organizations concerned with trade-related issues include the *World Trade Organization* (WTO) and the *World Intellectual Property Organization* (WIPO).

8.3   Industry organizations

Most countries have individual national associations that represent manufacturers, distributors and private pharmacies. These associations promote members' interests and provide support services such as training and dissemination of information. There are however a number of international organizations which represent different aspects of the industry. We consider them here.

8.3.1   The International Federation of Pharmaceutical Manufacturers Associations

The International Federation of Pharmaceutical Manufacturers Associations (IFPMA) is a non-profit NGO which represents 59 national industry organizations from both developed and developing countries. It exchanges information within the international industry and develops position statements on matters of policy. It is also the main channel of communication between the industry and various international organizations concerned with health and trade-related issues, including the WHO, the World Bank, the WTO and the WIPO.

8.3.2   The International Federation of Pharmaceutical Wholesalers

The International Federation of Pharmaceutical Wholesalers (IFPW) plays a similar role to the IFPMA, but with respect to pharmaceutical distributors.

8.3.3   The International Generic Pharmaceutical Alliance

The International Generic Pharmaceutical Alliance (IGPA) was founded in March 1997 and consists of the Canadian Generic Pharmaceutical Association (CGPA), the European Generic Medicines Association (EGA), the Generic Pharmaceutical Association (GPhA), and the Indian Pharmaceutical Association (IPA). It is the official representative body of the generic industry in relations with the ICH, WHO and other international organizations.

8.4   Other organizations

A wide range of NGOs plays a vital role in promoting consumers' and public health interests. In some countries, missions and other NGOs provide a substantial portion of health care and pharmaceutical service. The major NGOs involved in the management of pharmaceuticals in international health have already been discussed in chapter 7. However, one organization not mentioned in chapter 7 is the International Pharmaceutical Federation (FIP), founded in the Netherlands in 1912. FIP, as well as other organizations that play small but important roles in managing pharmaceuticals in international health, including government agencies and religious foundations, are discussed now briefly.

8.4.1   International Pharmaceutical Federation

FIP is a world-wide federation of national pharmaceutical (professional and scientific) associations. It connects, represents and serves over a million pharmacists and pharmaceutical scientists around the world. FIP has NGO status with the WHO. It aims to develop the role of pharmacists in the health care system and to promote the rational use of drugs.

8.4.2   Pharmacopoeia commissions

A number of national pharmacopoeial bodies make an important contribution to the management of pharmaceuticals in international health. For example, the *United States Pharmacopoeia* (USP) establishes state-of-the-art standards for more than 3,800 medicines, dietary supplements and other health care products. It is involved in developing and translating drug information materials and works towards harmonising international pharmacopoeial standards.

8.4.3   Faith-based organizations

Missions and other church organizations work with other NGOs in providing health care and pharmaceutical services in many countries. These are collectively known as Faith-Based Organizations, or FBOs. The *Churches' Action For Health/ Christian Medical Commission* (CMC) maintains drug distribution services in many of these countries. At international level, the CMC acts as a coordinating body and clearinghouse for information.

8.5   Essential drugs

The WHO defines essential drugs as "those that satisfy the needs of the majority of the population. They should therefore be available at all times, in adequate amounts, and in the appropriate dosage forms and at a price that individuals and the community can afford".

In most countries the selection of drugs is a two-step process. First, market approval is granted based on efficacy, safety and quality; then comparisons are made to limit procurement and/or reimbursements of drug costs, to identify the treatments which provide best value for money. The implementation of an Essential Drugs Programme (EDP) addresses the need for this second step.

8.5.1   The essential drugs concept in drug selection

The first WHO model list of essential drugs was devised in 1977 in response to the needs of developing countries. The affordability concept in the WHO's definition of essential drugs was introduced into the description in 1999. At present, an increasing number of developed countries also use key components of the essential drugs concept. Examples are the positive reimbursement list of the Pharmaceutical Benefits Scheme of Australia, the Scottish Intercollegiate Guidelines Network (SIGN) clinical guidelines, and some health maintenance organizations in the USA. In most cases this development was triggered by increasing drug costs and the introduction of many new and often expensive drugs which offered little advantage over previously available medicines. An essential drugs list is currently implemented in over 150 countries.

The WHO's selection criteria for its model list of essential drugs are relevance to the prevalence of diseases, proven efficacy and safety, evidence of performance in a variety of settings, adequate quality and stability, favourable cost-benefit ratio, a preference for well-known drugs with good pharmacokinetic properties, and for single compounds. Where drugs appear to be similar in the above respects, comparative pharmacokinetic properties, and availability of facilities for manufacture or storage are used as secondary criteria. Because of great differences between countries, the preparation of a drug list of uniform and general applicability is neither feasible nor possible. Countries evaluate and adopt their own lists of essential drugs, and revise them every two to four years.

The WHO Essential Drugs Committee is composed of experts in clinical medicine and pharmacology, pharmacy and clinical microbiology, and health-care workers. It reviews the WHO model list every two years, based on propositions submitted to the secretariat together with justification and references.

8.5.2 The evolution of the drug selection process

Major trends in health care have led to a more integrated approach to the development of essential drug lists and treatment guidelines, more emphasis on healthcare and public health perspectives, and have shown the need for regular monitoring of implementation and impact.

In the process of drug selection, there is a change from a normative, pharmacological approach to an epidemiological one, involving the identification of the burden of illness, drawing up a list of diseases and drafting accepted treatment guidelines based on best available scientific evidence.

8.5.3 Lists of essential drugs

Many developed countries have embarked on large-scale programmes to develop evidence-based standard clinical guidelines. The science of evidence-based decision-making has rapidly become the international norm. Increasingly, the strength of recommendations is being linked to the strength of the underlying evidence. The early institutional essential drugs lists were supply lists drawn up by national drug procurement agencies. Currently, decisions on national essential drugs lists are no longer taken in isolation and are increasingly subject to national clinical choices. The techniques of critical appraisal and pharmacoeconomics are used to draw up standard treatment guidelines and to derive an essential drug list from them.

Essential drug lists linked to standard treatment guidelines can guide procurement in the public sector, training and supervision, reimbursements, donations, local manufacture of drugs, and all aspects of the pharmaceutical system. The Thirteenth Model List (2003) contains 316 active substances. The list is presented on the internet (http://mednet3.who.int/eml/) with links to diseases, clinical indications and WHO clinical guidelines where they exist, as well as to the WHO Model Formulary.

8.5.4 Developing effective drug policies

The groups of organizations referred to above, and indeed individual organizations amongst them, have relationships that are often adversarial, due to their differing objectives. The international pharmaceutical industry, for example, is profit-driven, whilst many of the other organizations have the objective of providing safe, effective and appropriate drugs at lowest cost. The picture is further complicated by the fact that health care providers are not necessarily responsible for funding medicines, so that the funders or payers, with the objective of "value for money", add to an already complex situation.

Despite the increasing number of medicines on the market, one-third of the world's population lacks access to essential medicines; in the poorest parts of Africa and Asia this figure rises to one-half. Many factors influence access to effective medicines. On a local and national level they include the quality of diagnosis; rational prescribing, drug selection, distribution and dispensing of medicines; drug quality and capacities of health systems and budgets. Internationally, they include lack of R&D for treatment of "neglected" diseases, the abandonment of unprofitable but medically necessary drugs by the manufacturers; and price.

Improving access to drugs in poor countries requires major financing efforts. Whilst in middle income developing countries, some domestic resources can be mobilised, most of the additional financing will have to come from the international community. WHO has suggested the foundation of a new international health fund to redistribute resources through a fast, transparent and accountable process. Policy decisions and priority-setting would still be made at national level.

8.6   Pharmaceutical procurement

Pharmaceutical procurement is a crucial process in the management of pharmaceuticals, and involves knowledge of a range of important issues. These matters include pricing, marketing and promotion, international agreements and intellectual property matters, e-commerce, counterfeit drugs, and access to new drugs. These items will be discussed in the following sections.

8.6.1   Pricing

High prices and a lack of transparency in pricing strategies are major obstacles affecting access to essential medicines. Pharmaceutical companies, like all other market players, act according to economic, profit-maximising principles. They price their drugs according to what the market will bear. Pharmaceutical companies claim that the average cost of bringing one new medicine to market is US $ 500 million, that it takes an average of twelve to fifteen years to develop a new medicine, that for every 5,000 medicines initially evaluated, only five reach the stage of clinical trials and only one is finally approved for patient use. They state that while the cost of developing drugs is soaring, time to recover investments is shrinking because of stepped-up competition from generic drugs. The magnitude of these claims has been controversial.

R&D technology and costs evolve over time. Investments represent intangible capital and carry a high risk. To provide an incentive for continued R&D, drugs are protected by patent rights (see section 8.7). In recent times, particularly in the light of the worldwide HIV/AIDS epidemic and the lack of access to drugs in some of the worst affected countries, controversies have arisen around patent issues. Some discounts and donations have been granted to poor countries for vaccines, contraceptives and tuberculosis drugs. Generic competition has brought prices down to 10 per cent and less of the original price.

One way to improve access to drugs in poor countries, while maintaining incentives for R&D, is to apply differential pricing. Differential pricing is the adaptation of prices charged by the seller to the purchasing power of governments and households in different countries. Differential pricing is economically feasible, particularly where fixed costs – like the high R&D costs – are substantial, and variable or marginal production costs are low. Various ways of applying differential pricing have been suggested, including bilateral negotiations between companies and governments, regional or global bulk purchasing, moral suasion, and voluntary or compulsory licensing.

For a differential pricing system to work, markets must be effectively segmented to protect the interests of both consumers and manufacturers. Low-price products must not be diverted into rich countries, and consumers must be ready to accept price differences. The aim will be to identify the legal, institutional and political conditions that will be conducive to companies to engage in differential pricing, acting independently of each other and not as part of any concerted arrangement among themselves.

A number of controversial questions remain:

· Should a differential pricing system be market-regulated or globally imposed?
· What is the role of donations?
· What legal constraints are imposed by national competition laws (e.g. "anti-dumping" relief)?
· How should a "marginal cost" or "not-for-profit" price be determined?
· Is it possible to set "target prices" relating to therapeutic value through economic analysis?

Another approach to reduce drug prices is to stimulate competition by promoting generic drugs. This strategy will stimulate competition. Care must be taken to assure that generic drugs are of equal quality as branded products, and acceptance by professionals and public must be promoted.

Where no generic drugs are available, transparent competition should be encouraged for patented new drugs. Furthermore, prices can be contained by making use of the safeguarding provisions of the TRIPS agreement, such as compulsory licenses and parallel imports (see section 8.2.3).

8.6.2 Marketing and promotion

Medicines are different from other consumer products. The consumer, i.e. the patient, does not always choose the drug, it is prescribed for him by a health professional. Even if patients, or their parents, do choose their own medicine, they are not trained to judge its appropriateness, safety, quality or value for money. Fear of illness may lead patients to demand or buy costly drugs when cheaper medicines, or no medicines, would have achieved the same or even a better result. On the other hand, a patient often cannot judge the consequences of not obtaining a needed drug.

Despite the uncertainties associated with treatment of illnesses, the concepts used in marketing pharmaceuticals are similar to those used for marketing other products. Marketing can be defined as "the science of matching capabilities to market opportunities and constraints to maximise gain, by optimising for each market segment the mix of product, place, price and promotion." Marketing has been identified as both a barrier and a bridge to evidence-based health care. Unethical promotion of drugs can hinder the implementation of evidence-based principles. On the other hand, marketing is amongst the consistently effective interventions for promoting behavioural change among health professionals.

Inappropriate promotion of medicinal drugs remains a problem both in developing and developed countries. Promotion has been broadly defined by the WHO as "all informational and persuasive activities by manufacturers, the effect of which is to induce the prescription, supply, purchase and/or use of medicinal drugs".

Medicines are advertised to health professionals by many directly recognisable routes: in journals, at congresses, by direct mailing, by drug representatives, through the distribution of samples and other promotional material. Over-the-counter drugs are amply advertised to consumers in the media, while DTCA of prescription medicines is currently only allowed in two countries, the USA and New Zealand. However, prescription medicines are increasingly advertised worldwide via the internet. It has been claimed that drug promotion helps to achieve health benefits by disseminating information and empowering patients. However, many surveys have shown that advertisements commonly contain misleading and inaccurate information and are of poor educational value, as their primary aim is to promote sales.

As one element in the implementation of its 1985 Revised Drug Strategy, WHO developed a set of Ethical Criteria for Medicinal Drug Promotion, published in 1988 as a model for national guidelines for promotional practice. Nevertheless, two-thirds of the world's countries either have no laws to regulate pharmaceutical promotion or do not enforce the regulations they have. Studies in Belgium, UK and the USA have all shown a correlation between the use of information from sales representatives and inappropriate prescribing.

The IFPMA Code of Pharmaceutical Marketing Practices, a voluntary industry self-regulatory code, lacks any mechanisms for active monitoring, effective sanctions, or clear procedures to correct misleading information.

Drug promotion also takes place under the guise of continuing medical education, in promotional trials and through funding of research, or of health professionals' or patients' organizations. Clinical trials form the basis of effective research and development. While most clinical research is still conducted to high standards of objectivity, there is an increasing trend towards inappropriate involvement of sponsors, conflicts of interest on the part of investigators, and bias in publicising results. The WHO has therefore called for a new declaration, similar to the Helsinki Declaration drawn up to protect trial subjects, which would stem the tide of commercial influence to preserve the integrity of the clinical evidence base.

8.7   International agreements and intellectual property rights

The *WTO*, based in Geneva, deals with multi-lateral trade agreements on goods, services and intellectual property. Its main underlying principles are non-discrimination, transparency and predictability of international trade. The WTO agreements most relevant for the pharmaceutical sector are the *Agreement on Technical Barriers to Trade* (TBT), the *General Agreement on Trade in Services* (GATS), and the *Agreement on Trade-Related Aspects of Intellectual Property Rights* (TRIPS).

8.7.1   Agreement on Technical Barriers to Trade

The TBT encourages countries "to base technical regulations ... on international standards ... when these exist or their completion is imminent". In the context of pharmaceuticals, international standards for drug development and test procedures for Europe, Japan and the USA have been set by the *ICH*, as part of a broader quality assurance system. It remains to be seen whether they should become global standards, and if not, what the alternatives are.

8.7.2   General Agreement on Trade in Services

The GATS defines restrictions on a broad range of government measures that may affect the trade in services. In a pharmaceutical context, it concerns international (foreign) investment, foreign patients, overseas treatment, and the influx/outflux of health personnel.

8.7.3   Agreement on TRIPS

The most important area covered by international agreements in the field of pharmaceuticals is that of the protection of intellectual property rights. Many developed countries introduced pharmaceutical patents in the 1960s or 1970s, when their industries had reached a certain degree of development. Many developing countries did not begin to grant patent protection for pharmaceutical products until the late 1980s, though most did recognise process patents. Facing the threat of trade retaliation under the negotiations for the General Agreement on Tariffs and Trade (GATT) due to what was considered to be their lack of or inadequate protection for pharmaceuticals, many developing countries changed their laws accordingly: these included Chile, Mexico and South Korea.

In 1995, the *Agreement on TRIPS* came into force. It is implemented in cooperation with the WIPO. It covers copyright, patents, trademarks, trade secrets, and related issues in all fields of technology, and concerns both product and process inventions. Large pharmaceutical firms attach special value to this protection because of the considerable expenditure involved in the development of new drugs, and the fact that new products may be imitated relatively easily, as suggested by short imitation time-lags. The standard term of protection for patents is 20 years. TRIPS is binding on all WTO member states. Dissident states can be disciplined through the WTO's dispute settlement mechanism, which can authorise trade sanctions against them.

Developing countries that did not provide product patent protection at the time TRIPS came into force in 1995 had a transitional period of up to 11 years to introduce the protection. This period was subsequently extended until 2015 for least developed countries.

TRIPS standards allow some flexibility regarding the criteria for patentability. If these criteria are too broad, they will allow "evergreening", i.e. re-patenting of a drug with a slight modification when the original patent protection expires, and thus reduce or prevent access due to continuing high prices.

8.7.4   Impact of the TRIPS agreement

In the pharmaceutical field, the implementation of TRIPS standards delays the introduction of new generic products, resulting in increased drug prices and therefore reduced access. Even after a patent expires and competition from generic products develops, the brand product may still maintain a high price due to brand loyalty. Such a situation may have a negative impact on the pharmaceutical industry in developing countries.

Proponents expect TRIPS to enhance local innovation, and to increase foreign direct investment and technology transfer. However, a study in Thailand, where pharmaceutical patents were introduced in 1992, showed that this was not the case. In October 1999 the Centre for Health Economics at Chulalongkorn University, published a study of "The Implications of the WTO TRIPS Agreement for the Pharmaceutical Industry in Thailand". They concluded that "it seems that more negotiation rounds and revisions of international agreements have maximized the exclusive rights of patentees, but minimized the rights of people who need innovations". They also argue that there has not been much technology transfer or foreign investment directly into the pharmaceutical industry in Thailand since 1992.

Given the substantial finance needed to develop a new drug, very few developing countries' firms can support the necessary R&D expenditures. This situation is particularly the case in the sector of biotechnology, where drugs are modelled on natural substances, and no new drugs can be developed based on similar compositions. Most product patents in developing countries are held by foreign enterprises. As a result, national industrial development may be substantially hindered, and repatriated profits and royalties will have an impact on the balance of payments.

8.7.5   TRIPS safeguards

The TRIPS agreement contains the following *safeguards* for countries to counteract these negative effects.

· *Compulsory licenses* can be granted without the permission of the patent holder for reasons of public health, national emergency or extreme urgency, public non-commercial use, to remedy anti-competitive practices. TRIPS does not limit the grounds for issuing a Compulsory License. However, TRIPS does impose strict conditions on Compulsory Licensing, e.g. case-by-case decisions must be taken, attempts to obtain a voluntary licence must have failed, the patent holder must be "adequately" remunerated, and the product must be destined predominantly for the domestic market.
· *Parallel importation*, i.e. importation without the patent holder's consent, of a patented product marketed in another country. Article 6 of the TRIPS Agreement states that practices relating to parallel importation cannot be challenged under the WTO disputes settlement system, provided that there is no discrimination on the basis of the nationality of the persons involved.
· *Provision for early working* ("Bolar provision") allows testing and regulatory approval of generic drugs before the patent protection of the original drug expires. This provision facilitates generic competition.

8.7.6 Implementation of TRIPS

TRIPS is implemented through national law. It is crucial that countries design national legislation which allows them to protect the public health interest. Thus, health officials must take part in the discussion and provide input when national laws are designed.

The agreement has received a growing level of criticism from developing countries, academics and NGOs. The most visible conflict has been over AIDS drugs in Africa. This controversy has not led to any revisions to TRIPS. Instead, responding to demands from developing countries, the WTO meeting at the Fourth Ministerial Conference in Doha, Qatar, adopted a declaration on TRIPS and public health in November 2001.

The *Doha Declaration* indicated that the TRIPS agreement should not prevent members from dealing with public health crises by using the agreements safeguarding provisions, that solutions should be investigated to assist countries lacking manufacturing capacities to make use of the compulsory licensing provision, and that technology transfer should be encouraged. The transitional implementation period for least-developed countries was extended to 2015.

Since that time, at the behest of PhRMA (Pharmaceutical Research and Manufacturers of America), the United States and, to a lesser extent, other developed nations have been working to minimise the effect of the declaration. From an industry perspective, company representatives have encouraged governments to accept offers for drug donations and discounted pricing.

8.8 Case study-HIV/AIDS and antiretroviral treatment in South Africa

We conclude this chapter with a brief description of the situation in South Africa with regard to HIV/AIDS and antiretroviral treatment. It was a situation which achieved international prominence and extensive coverage in the news media. In many ways it came to symbolise the key issues concerning the management of pharmaceuticals in international health during the 1990s and early 2000s, and illustrates the political nature of the interaction between the main stakeholders.

8.8.1   Origins and initial responses

The first cases of HIV were diagnosed in South Africa in the early 1980s. Subsequently there was an explosion in HIV prevalence between 1993 and 2000, at a time when the country was distracted by major political and social changes. Efforts to procure antiretroviral drugs to treat people living with HIV/AIDS were impeded by international pressure to protect patents on antiretroviral drugs held by multinational pharmaceutical companies, as well as lack of political leadership at the national level.

The first governmental response to AIDS came in 1992 when Nelson Mandela addressed the newly-formed National AIDS Convention of South Africa (NACOSA). The prevalence rate, based on antenatal testing, was 2.4 per cent.

In 1996 South Africa developed a national drug policy, intended to make essential medicines available to all citizens who needed them. The policy recommended a number of strategies to meet these objectives, including legal and regulatory mechanisms encouraging generic substitution of medicines not under patent, parallel importation of medicines and purchasing of generic medicines on the international market. The prevalence rate of HIV, based on antenatal testing, was now 14.2 per cent.

The Medicines Act of 1997 mandated three important initiatives: it authorized compulsory licensing (whereby states authorize generic production of a patented product without the patent holder's consent); it authorized parallel importation (cross-border trade in a product without permission of the patent holder); and it promoted the practice of prescribing less expensive generic versions of patented drugs, including antiretroviral drugs essential in fighting HIV/AIDS. The Medicines Act however did not come into force.

8.8.2   Legal action by the industry

In 1998, the South African Pharmaceutical Manufacturers Association and forty-one multinational drug companies filed an action in the Pretoria High Court to prevent the government from bringing into operation certain sections of the Act that would allow the government to produce or import drugs at lower cost. Intense lobbying efforts by multinational pharmaceutical companies and the United States government further undermined South Africa's capacity to implement the provisions of the Act. The USA's efforts included threatened trade sanctions.

In 1998 the Treatment Action Campaign (TAC), a non-governmental organization with a nationwide base was founded to advocate for the rights of people living with HIV/AIDS. The then Deputy President Thabo Mbeki launched the Partnership Against AIDS. The prevalence rate was now 22.8 per cent based on antenatal testing.

In 1999 the TAC coordinated an aggressive international campaign urging the pharmaceutical industry to lower the prices of patented drugs and calling for the United States government to halt its efforts to roll back the Medicines Act. In September 1999, following intense activist pressure in both South Africa and the U.S., the U.S. government conceded the validity of the Medicines Act. In December 1999, it removed South Africa from its list of countries identified as lacking intellectual property rights protection (the "301 Watch List").

In 2000, President Mbeki set up a presidential AIDS advisory panel. This panel included HIV "dissidents" who believe that AIDS in Africa is not caused by HIV, but by poverty, poverty-related conditions and illnesses, and medicines used to treat HIV.

8.8.3 Withdrawal of legal action

In April 2001 the Pharmaceutical Manufacturers Association withdrew its court case. In the same year, South Africa's High Court ordered the government to make nevirapine available to prevent mother-to-child transmission of HIV. Despite international drug companies offering free or cheap antiretroviral medicines, the Health Ministry failed to provide these drugs on a large scale.

In 2002 the prevalence rate was 26.5 per cent based on antenatal testing. TAC campaigners embarked on a strategy of civil disobedience and demonstrations to try to embarrass the government into acting. In March 2003 TAC laid culpable homicide charges against the Ministers of Health and of Trade and Industry.

In August 2003, the government ordered the health department to develop a detailed operational plan to provide antiretroviral drugs. In October 2003 the Clinton Foundation announced that it had brokered a deal with four generics companies to provide triple-drug antiretroviral therapy to governments in the developing world at a cost of less than US$ 140 per patient per year. Glaxo SmithKline and other pharmaceutical companies agreed to allow low-cost generic versions of their drugs to be produced. In November 2003, the government approved the Operational Plan for Comprehensive Care and Treatment for people living with HIV and AIDS.

In February 2004, the government in South Africa stated that delays in the procurement process and lack of training and staff shortages in the health care system were still delaying the rollout of ARV treatment.

Note

The principal sources for this case study were: Human Rights Watch. Deadly Delay: South Africa's Efforts to Prevent HIV in Survivors of Sexual Violence. March 2004.

Vol. 16, No. 3 (A). Available online at http://www.hrw.org/reports/2004/ southafrica0304/4.htm#_Toc65295177, accessed 9 March 2004. AVERT website. HIV and AIDS in South Africa. Available on line at http://www.avert.org/ aidssouthafrica.htm, accessed 9 March 2004.

8.9   Conclusion

In this chapter we have discussed some of the central issues in relation to access to pharmaceuticals by the people of developing countries. We considered each of the major international players, particularly the industry and WHO. We have focused on the core issue of essential drugs and their pricing and procurement. And we have reviewed the mechanisms, by way of international agreements, through which these issues are debated and agreed.

It is clear from this discussion that the issues themselves are very complex, and that power between the various actors is shared unequally. Nevertheless, existing agreements represent considerable progress, despite the fact that this has been slow and difficult. There remains an enormous amount still to be done, and the challenges for the future are to make further progress more rapidly by fully engaging all those involved.

Further reading

Chang, H.-J. (2002) *Kicking Away the Ladder-Development Strategy in Historical Perspective*. London: Anthem Press.

Grace C. (2004) *Equitable Pricing of Newer Essential Medicines for Developing Countries: Evidence for the Potential of Different Mechanisms*. http://www.who.int/medicines/library/par/equitable_pricing.doc.

Laing, R., Waning, B., Gray, A., Ford, N. and t'Hoen, E. (2003) "25 years of the WHO essential medicines lists: progress and challenges". *Lancet* 361: 1723–1729.

Pécoud, B., Chirac, P., Trouiller, P. and Finel, J. (1999) "Access to essential drugs in poor countries: a lost battle?" *Journal of the American Medical Association* 281: 361–367.

Reich, M.R. (2000) "The global drug gap". *Science* 287 (5460) :1979–1981.

Velasquez, G. and Boulet, P. (1999) *Globalization and Access to Drugs. The impacts of Intellectual Property Rights on Access to Medicines*. Health Economics and Drugs DAP Series, No.7; WHO/DAP/98.9 Revised. Geneva: World Health Organization.

WHO (2003) *Intellectual Property Rights, Innovation and Public Health. Report by the Secretariat*. Geneva: World Health Organization. http://www.who.int/gb/ EB_WHA/PDF/WHA56/ea5617.pdf

Chapter 9
Rational Use of Medicines

Karin Wiedenmayer

Box 9.1: Learning objectives for chapter 9

By the end of this chapter you should be able to:

· Describe the main consequences of the non-rational use of medicines.
· Describe non-pharmacological factors that determine medicine use.
· List the three main types of indicators for medicine use.
· Explain the main features of educational, managerial and regulatory approaches to medicine use intervention.
· List three interventions to promote rational medicine use known to be effective and three known to be ineffective.
· Describe the roles of prescribers and dispensers in promoting rational medicine use.
· Describe the role of patients and public education in promoting rational medicine use.

9.1   Introduction

In chapter 8 we described the concept of essential medicines, introduced by WHO to focus attention on action to increase access to those medicines that make the greatest difference to the health status of a population. We noted that the concept of rational medicine use has been accepted as a major priority in many countries. In this chapter we consider the consequences of non-rational medicine use, explore medicine use behaviour in greater detail, describe ways of investigating medicine use problems, and review medicine use interventions and methods of evaluating their effectiveness.

Irrational prescribing, dispensing and patient use of medicines is a worldwide problem. Up to 75 per cent of antibiotics are prescribed inappropriately, even in teaching hospitals, worldwide. Antibiotic resistance is growing rapidly for most major infectious diseases, largely as a result of non-rational use. Typically, only 50 per cent of patients take their medicines correctly, worldwide. The actual use of medication often takes little account of therapeutic principles based on sound clinical data.

Initiatives to improve matters have been taking place for a number of years. In 1985 the Nairobi Conference of Experts highlighted the importance of rational medicine use by prescribers, dispensers and patients. They acknowledged that available medicines may be wasted if not used rationally.

In 1997, researchers and policy makers from around the world gathered in Chiang Mai, Thailand, for the first International Conference on Improving the Use of Medicines (ICIUM). The conference was a watershed for policy in this area. It focused attention on cost-effective and innovative interventions for use in developing countries. It also defined a new global research agenda relevant to current conditions and unfolding developments in international health. A second ICIUM conference will be held in 2004.

In recent years, the field of international health has experienced advances as well as setbacks. The potential benefits of varied health reforms, decentralization, a fast-growing private sector, and new medicine financing and incentive schemes have been explored in many settings. Several new global initiatives have arisen to address catastrophic epidemics and improve access to essential medicines, especially antimicrobials. However, the increased global flow of antimicrobials brings with it the twin threats of growing rates of antimicrobial resistance and rising prices for alternative medicines.

9.2   Consequences of non-rational medicine use

The costs and impact of non-rational medicine use are enormous. They include inefficient use of limited resources, adverse clinical consequences, and unnecessary suffering of patients. It can lead to a reduction in the quality of medicine therapy, causing increased morbidity and mortality. Waste of resources leads to increased costs and possibly overuse of other vital medicines. There is increased risk of unwanted effects such as adverse drug reactions and the emergence of medicine resistance. There is a psychosocial impact for patients that may lead to the belief that there is "a pill for every ill". Studies in both developed and developing countries describe numerous examples of non-rational medicine use. Box 9.2 illustrates some of them.

The issues of adverse drug reactions and the emergence of antibiotic resistance are explored in chapter 10. These are examples of serious consequences of non-rational use of medicines, and the figures suggest that the situation is continuing to deteriorate. The shift in many countries from public to private sector funding of health care has tended to make matters even worse. A combination of business interests and ignorance contributes to the ever more indiscriminate and irrational use of medicines, especially in countries without proper medicine regulation.

9.2.1   Definition of rational medicine use

WHO defines rational medicine use in the following terms:

· The rational use of medicines requires that patients receive medications appropriate to their clinical needs, in doses that meet their own requirements,

---

Box 9.2: Examples of non-rational medicine use

· Polypharmacy (multiple or over-prescription), use of medicines that are not related to diagnosis.
· Unnecessarily costly medicines.
· Inappropriate use of antibiotics.
· Indiscriminate use of injections.
· Irrational self-medication with under-dosing and over-dosing.
· Incorrect administration, dosages, duration.
· The use of medicines when no medicine therapy is indicated.
· Failure to prescribe available, safe and effective medicines.
· Poor compliance with tuberculosis treatment.
· Under use of effective medicines for hypertension and depression.
· Hospital medicine use problems such as antibiotic misuse for surgical prophylaxis.

---

for an adequate period of time and at the lowest cost to them and their community.

The WHO definition is expressed mainly in medical and financial terms. But, as we have seen in chapter 3, people have their own rationale for deciding on therapies. Irrationality defined from a medical point of view may be totally rational from the consumer's point of view. Rational medicine use cannot only be defined by the universal criteria of safety, efficacy, affordability and need.

9.2.2   The process of medicine use

Medicine use has to be seen in both biomedical and local social contexts. It encompasses the main actors involved. These are prescribers, dispensers and patients. The process entails

· *diagnosis* (identifying what is wrong),
· *therapy* (therapeutic objective and plan),
· *prescribing* (information, instructions, warnings and prescription writing),
· *dispensing* (supply, advice, counselling, instructions and warnings) and
· *adherence* by the patient (understanding of therapy, patient responsibility and valuing of treatment).

9.3   Medicine use behaviour

Individual medicine use patterns occur within a medicine use system as part of a larger health care system. This in turn is influenced by the social, cultural, economic and political context of the country concerned. Medicine use also takes place within a network of beliefs and motivations on the part of the provider and the patient.

9.3.1   The non-pharmacological basis of therapeutics

It might seem reasonable to suppose that medicines are prescribed, dispensed and taken in order to prevent, alleviate or cure illness. This is essentially true. However, other factors influence patterns of medicine usage and illness, and these may be unrelated to the actual properties of the medicines being taken. These factors have been called the *"non-pharmacological basis of therapeutics"*. Patterns of medicine consumption in a country will therefore reflect not only its scientific and technological resources but also its socio-demographic state, its traditions and its culture.

In practice, rational medicine use implies that the observed prescribing behaviour should be compared to an agreed norm or standard. Medicine use cannot be studied without a method of measurement and a reference standard. This leads to a scientific biomedical model of pharmacotherapy. But the psycho-social setting in which a medicine is given or taken may also influence people's response to the medicine, as illustrated by the placebo effect. The placebo effect is generally culture-specific, since it depends on how people view the world. It can be modified by factors such as the social setting in which the treatment is administered. Cultural factors pervade all aspects of medicine use, which is embedded in a matrix of social values and expectations.

9.3.2   The cultural perspective

The cultural setting determines how society views the use of medicines, what its social acceptability is, and what social significance it attaches to it. The importance of the cultural perspective is highlighted in anthropological studies of medicine use in less developed countries. Some of these studies illustrate how cultural definitions of health and illness lead to a reclassification or reinterpretation of Western medicines. Failure to understand the local cultural perspective of medication will lead to inadequate communication about medicine use. People base their actions on what they believe.

To understand medicine use we have to learn from different paradigms. Early efforts to improve medicine use tended to assume that irrational use was largely due to lack of knowledge, and that the simple solution was to provide information. But it soon became clear that the reasons for irrational medicine use were complex and multi-factorial, and included a mass of social and cultural factors, economic incentives and promotional practices. Complex problems demand complex strategies, and different disciplines can contribute to an understanding of patterns of medicine use. It becomes clear that a multidisciplinary approach is necessary to tackle problems of irrational medicine use.

9.4   Investigating medicine use problems

Only by having a thorough understanding of three key issues can we attempt to improve the quality and efficiency of pharmacotherapy. The issues are:

- existing patterns of therapy,
- the magnitude of departure of these therapies from optimal practice, and
- the factors (clinical, psychological, political, economic and cultural) which determine them.

9.4.1   Factors contributing to non-rational medicine use

Factors underlying irrational medicine use are as multi-factorial as the medicine use system itself. As with other processes medicine use has first to be examined and assessed quantitatively and qualitatively before an appropriate intervention can be planned. "Medicine use encounters" can be defined as the interaction between a service provider and a patient at which decisions are made about which medicines to recommend or use. Medicine use encounters take place in many different locations, such as the hospital, private practice, a pharmacy, at home, health centre, dispensary, at the traditional healer, with a medicine seller or on the marketplace.

In order to measure medicine use behaviour we need to collect data. Both qualitative and quantitative methods should be used to identify the motivations and incentives of prescribers, dispenser and patients. The method chosen to study medicine use depends on the nature of the problem, the objectives of the study, resource and time availability, local capacity and experience. It is best to use multiple methods and to triangulate findings, since each method can look at different aspects of a problem.

9.4.2   Medicine use indicators

WHO and INRUD have developed indicators of medicine use to measure and assess medicine use patterns. Undertaking a medicine use indicator study is possible in nearly all environments. Indicator studies can be used for descriptive and comparative studies, to monitor performance and to evaluate interventions. They include indicators of prescribing, patient care and facilities. Key indicators are illustrated in Box 9.3.

Medicine use indicator studies are easy to conduct and do not need extensive research experience. The WHO Guide leads the practitioner through all the necessary steps that have to be taken. Indicator forms are simple and can be completed manually or electronically.

Box 9.3: Medicine use indicators

Prescribing indicators:	Average number of medicines Percentage of antibiotics Percentage of injections Percentage of generics Percentage prescribed from national essential medicine list or formulary
Patient care indicators:	Average consultation time Average dispensing time Percent medicines dispensed Per cent medicines adequately labelled Patient's knowledge of correct dosage
Facility indicators:	Availability of national essential medicine list or formulary Availability of key medicines

Source: How to Investigate Drug Use in Health Facilities, Selected Drug Use Indicators (1993) INRUD/WHO: International Network for Rational Use of Drugs www.msh.org/inrud/activities.html

---

9.5   Medicine use interventions

Once a situation analysis has been conducted, interventions can be designed. Four main strategies have been developed to improve medicine use, ranging from simply providing information to restrictive regulatory measures. The strategies are *educational* (to inform or convince), *managerial* or *administrative* (to guide practice), *regulatory* (to restrict choices) and *economic* (to offer incentives).

In general, medicine use can be improved through interventions targeted at specific problems with an identifiable audience (for example, inappropriate antibiotic prescribing to treat urinary tract infections by private physicians); or by a system's change looking at broader problems and more diffuse target groups (for example, the unnecessary prescribing of expensive branded medicines).

9.5.1   Educational approaches

Educational approaches seek to inform or persuade prescribers, dispensers or patients to use medicines in a different way, by means of information, knowledge or persuasion. Printed materials are the most common and least expensive form of educational intervention. They can include scientific literature, pharmacy and therapeutics newsletters, and printed guidelines. Using printed materials alone as a means of improving prescribing makes two assumptions: firstly, that the main reason for incorrect prescribing is lack of information; and secondly that if prescribers had the correct information, their prescribing would improve.

However, this is not always the case; studies have shown that distributing printed material alone results only in brief, very small or even no improvements in prescribing. Often these materials are not even read. On the other hand, printed materials, particularly ones that are well constructed with easy-to-read messages, are an essential part of a total programme that also includes more intensive and individualized education.

Talking directly to practising prescribers about appropriate medicine use is a common and effective intervention strategy. Small group seminars can be successful if they are well focused. Interactive, participatory group discussions, clearly targeted, have been shown to have significant impact on medicine use.

9.5.2 Managerial and administrative approaches

Managerial approaches structure or guide decisions through the use of specific processes. The concept is to guide the practice of health practitioners. Infrastructure and the health care system are targeted and structured to influence medicine use.

Essential medicine lists, usually in the public sector, and national or institutional formularies, provide prescribers with a list of the medicines considered to be most effective, safe and economical in treating important health problems.

There have been few objective studies of the impact of medicine lists and formularies in developing countries. Whilst they probably reduce the use of unnecessary medicines by reducing their availability, it is also likely that without prescriber or patient education, other medicines continue to be misused. An example of the essential medicine concept is the medicine supply kit, where a limited number of medicines is supplied in fixed quantities at a regular interval to health facilities. A study has shown that medicine kits combined with training can significantly change medicine use indicators.

Standardized diagnostic and treatment protocols are decision rules which lead health workers to the most appropriate action based on patient symptoms and clinical signs. Therapeutic guidelines are a common method for disseminating standards of practice. Many countries have produced collections of guidelines that detail the preferred treatment for major health problems. A number of factors are important in determining how effective such guidelines will be in changing behaviour in different settings. These include having a participatory process with targeted prescribers, dissemination coupled with training, and follow-up with supervision.

9.5.3 Regulatory approaches

Regulatory approaches restrict allowable decisions by placing absolute limits on the availability of medicines. These strategies rely on rules or regulations to change behaviour. They are intended to restrict decisions rather than to simply guide

---

Box 9.4: Effectiveness of interventions

Interventions with proven effectiveness:	Face-to face education focused on few prescribing problems
	Problem-based professional basic training
	Prescription audits/medicine consumption review plus feedback
	Essential medicine lists and/or standard treatment guidelines developed by peer groups and followed by training
	Regular effectively designed training and followed by supervision
Interventions proven to be ineffective:	Printed materials alone
	Arbitrary limits on number/quantity of medicines per prescription
	Unfocused education
	Essential medicine lists or standard treatment guidelines alone

Source: Adapted from Laing, R., Hogerzeil, H., and Ross-Degnan, D. (2001) "Ten recommendations to improve use of medicines in developing countries". *Health Policy and Planning* 16 (1):13–20.

---

them, and are therefore usually designed to be inflexible. For regulations to be effective, however, there must be mechanisms for enforcement. Controls can be initiated by restricting public sector availability through the use of an Essential Medicine List (EML) or formulary. When there are major barriers to the prescribing of medicines not on an EML or formulary, this strategy takes on a regulatory character.

Another form of regulation is to encourage the use of generic, non-branded medicines. This can be done by limiting EMLs and formularies to generic forms of a medicine when available, requiring prescribers to prescribe by generic name rather than brand, and by allowing pharmacists to dispense the generic equivalent for a branded medicine prescribed by a physician. As with other types of regulatory measures generic policies can cause shifts in utilization to the private sector.

9.5.4   Economic approaches

Economic interventions target providers by offering incentives. By influencing medicine budgets providers may be convinced to change irrational medicine use. Consumers, on the other hand, can be made more responsible regarding medicine use with cost-sharing schemes. It will normally be in the consumer's interest to ask for generic medicines. These are more likely to be included in a formulary, and are also likely to be more affordable than branded products.

9.6 Effectiveness of interventions

Many studies have been carried out on interventions designed to influence medicine use, and the long term effectiveness of various medicine utilization interventions can be summarized. They are illustrated in Box 9.4.

The most effective medicine use intervention is one that is participatory, interactive, problem-based and focused. This includes targeted efforts, combined strategies, repeated and continuous training and supportive supervision. In general it is more efficient to combine a number of strategies to a multifaceted intervention to improve medicine use. To measure the impact of an intervention it is necessary to assess actual performance rather than knowledge. Knowledge by itself does not change behaviour: skills and performance is a better predictor of change.

WHO and the INRUD have extensive resources and literature on the promotion of rational medicine use. The INRUD Drug Use Bibliography is an annotated bibliography of published and unpublished articles, books, reports, and other documents related to medicine use, with a special focus on developing countries. It now contains over 5,000 entries and is updated regularly. The First ICIUM produced an expert international consensus on interventions to improve medicine use.

9.6.1 The role of the prescriber and dispenser

Both prescribers and dispensers play an important role in promoting rational medicine use. The classical role of a professionally qualified prescriber in promoting rational medicine use is diagnosis, prescribing and the provision of medicine information for the patient. In reality prescribers may be medical doctors, medical officers, nurses, nurse assistants and others, as we have seen in chapter 4.

The role of the pharmacy professional or dispenser is medicine supply and dispensing, medicine information for the patient, communication with prescriber and patient, and consumer education. Often dispensers include both pharmacists and pharmacy technicians, as well as nurses, nurses aids, doctors, medicine sellers, shop keepers and family members.

This suggests that traditional academic training for professionals may not be sufficient to improve the knowledge, skills and attitudes of health providers regarding the use of medicines and pharmacotherapy.

9.6.2 The role of the patient and public education

Patients have a central role in the use of medicines. They are the final recipients of health care, and they can significantly influence medicine use. Patients have their own concepts of health, disease aetiology, cure and medicines. Various factors influence medicine use by consumers or patients. People also use medicines according to their own ideas about efficacy and safety of medicines. Box 9.5 provides statements that illustrate some of the ways in which beliefs influence the need for medicines.

> **Box 9.5: How beliefs influence need**
>
> ---
>
> The following statements were made by people living in rural villages within the Karakoram Mountains in Pakistan:
>
> - "Medicine is needed for every illness. If medicine is not used, the illness will become serious."
> - "All illnesses need medicine. No illness will be cured without medicine."
> - "Medicine is to the sick, what water is to the thirsty."
> - "If we don't get medicine, how will we get cured?"
>
> ---
>
> *Source*: Rasmussen, Z.A. et al. (1996) "Enhancing appropriate medicine use in the Karakoram mountains". *Community Drug Use Studies*. Amsterdam: Het Spinhuis.

People do not only take medicines to treat symptoms of ill health. They also increasingly believe that medicines are needed to stay healthy or prevent illness. Preventive use of medicines is a topic often neglected in discussions on appropriate medicine use. However, medicine sales increasingly tend to involve such products. Vitamins are the most common type of preventive medicines.

There are several steps and decisions a patient goes through before actually taking a medicine.

- The patient must believe something is wrong.
- The patient must decide that it is serious enough to do something.
- The patient must choose where to seek help.
- The patient with a prescription must decide whether to buy medicines.
- The patient must decide whether to take medicines.

Patients also value medicines according to a range of criteria such as mode of administration and packaging, speed of action, cost, taste and colour, the source of the medicine, causes attributed to an illness and its perceived severity, their past experience with a certain medicine, the benefits of new as opposed to old medicines, and the effect of promotion. These ideas ultimately help shape medicine use practices. Some of the ways in which beliefs influence the use of medicines are illustrated in Box 9.6.

Very often the impact of culture on the use of medicines is mediated through gender. In many societies women take the central role in all matters concerned with health and disease. Box 9.7 gives examples of the influence of culture on women's role in health in the Philippines and in Pakistan.

9.7   Conclusion

In this chapter we have explored the main issues concerning rational use of medicines. We have described the main consequences of the non-rational use of medi-

---

Box 9.6: How beliefs influence use

In Sierra Leone, it was found that medicines' efficacy is linked to colour symbolism. Red medicines, for example, are thought to be good for the blood.

Source: Bledsoe, C.H. and Goubaud, M.F. (1985) "The Reinterpretation of western pharmaceuticals among the Mende of Sierre Leone". *Social Science and Medicine* 21 (3): 275–282.

In Uganda, one study describes the popularity of injections. People believe that medicine injected into the bloodstream does not leave the body as quickly as that administered orally. Oral medicine is compared to food, which enters the digestive system and eventually leaves the body through defecation.

Source: Birungi, H. et al. (1994) *Injection Use and Practices in Uganda*. WHO/DAP/94.18. Geneva: World Health Organization.

An investigation *in Ghana* describes how people consider heat to be the main cause of measles. Heat causes constipation and stomach sores in children. To treat measles people give Septrin (co-trimoxazole) syrup, multivite syrup, calamine lotion, akpeteshie (local gin) and a herbal concoction given as an enema to 'flush out' the heat.

Source: Senah, K.A. (1997) "The popularity of medicines in a rural Ghanaian community". *Community Drug Use Studies*. Amsterdam: Het Spinhuis.

---

cines, and demonstrated how failure to use medicines rationally undermines efforts to achieve more equitable access to them. We have considered the impact of medicine use behaviour, and explored methods for investigating problems that result. And we have identified a number of interventions that can be successful in changing medicine use behaviour.

Simply improving access to pharmaceuticals will not of itself solve the problems of the world's poor in securing adequate health care. As we have seen, all elements of the pharmaceutical supply chain need to function properly, so that medicines reach the people who need them wherever they are. Once they reach the patient we need to be confident about the quality of the medicines supplied. We also need to be aware of the consequences of the irrational use of medicines. These are most evident in the incidence of adverse drug reactions, and the development of resistance to antibiotics. We explore these issues in the next chapter.

Further reading

Dukes, M.N.G. (1993) *Drug Utilization Studies: Methods and Uses*. Geneva: World Health Organization.

Box 9.7: How culture influences women's roles

In the Philippines mothers decide whether or not they should buy medicines and give them to their children. Men are usually not involved in decision-making on the treatment of common childhood illnesses. Instead, women consult with neighbours and relatives on treatment options. Mothers and wives in this country manage household expenses and the family's income. They don't have to consult their husbands about costs. Husbands take a more active role only when a health problem becomes severe.

Source: Hardon, A. (1991) Confronting Ill health: Medicines, Self-care and the Poor in Manila. Quezon-City: Health Action Information Network.

In Pakistan women are constrained in their efforts to treat children's health problems. They cannot go to the bazaar or hospital in town to obtain medicines, as local, cultural norms forbid such mobility for women. For this reason, husbands, sons or other family members must buy medicines. As a consequence of these gender roles, men in this country are involved in decisions about children's treatment. They often receive information on a medicine's use at the bazaar or health facility and tell this information to their wives who actually administer the drugs.

Source: Rasmussen, Z.A. et al. (1996) "Enhancing appropriate medicine use in the Karakoram mountains". *Community Drug Use Studies*. Amsterdam, Het Spinhuis.

Essential Drugs Monitor 23 (1997) Geneva: World Health Organization.

How to Investigate Drug Use in Health Facilities, Selected Drug Use Indicators. (1993) INRUD/WHO: International Network for Rational Use of Drugs www.msh.org/inrud/activities.html

ICIUM: *International Conference on Improving Use of Medicines* (2004). http://www.who.int/medicines/organization/par/icium/icium.shtml

Kanji, N. Hardon, A., Harnmeijer, J.W., Mamdani, M. and Walt, G. (1992) *Drugs Policy in Developing Countries*. London: Zed Books.

Laing, R., Hogerzeil, H. and Ross-Degnan, D. (2001) "Ten recommendations to improve use of medicines in developing countries". *Health Policy and Planning* 16 (1): 13–20.

Promoting Rational Drugs Use Course (1994) Accra, Ghana: Management Sciences for Health, INRUD, WHO.

Promoting Rational Drug Use, MS Word and MS PowerPoint Course Materials available at http://dcc2.bumc.bu.edu/prdu/Word_Powerpoint_Files_TOC.html

Chapter 10
Medicine Quality, Adverse Reactions and Antimicrobial Resistance

Karin Wiedenmayer

Box 10.1: Learning objectives for chapter 10

By the end of this chapter you should be able to:

· Explain the difference between quality assurance and quality control.
· List the circumstances where quality control efforts should be targeted.
· List categories of medicines on which quality assurance efforts should be focussed.
· Describe key elements of a quality assurance programme.
· Define a counterfeit medicine.
· Describe factors contributing to the proliferation of counterfeit medicines.
· Define an adverse drug reaction and explain the difference between Type A and Type B reactions.
· Describe the steps that should be taken to assess possible adverse drug reactions.
· List the types of adverse drug reaction that should be reported.
· Describe factors contributing to antimicrobial resistance.
· Explain consequences of antimicrobial resistance.
· List strategies that can be used to contain antimicrobial resistance.

10.1   Introduction

Medicine quality can be compromised at all stages of the pharmaceutical management cycle, from the quality of the raw ingredients, through the processes of manufacturing, packaging, transport, storage, distribution and dispensing, to the eventual use of the medicine by the individual patient. Inferior medicine quality has effects at many levels. It poses a health risk for individual patients by leading to poor treatment outcomes and treatment failure, potentially causing toxic or adverse reactions, prolonged illness or even death. Substandard medicines or vaccines can cause resurgence of diseases or the development of resistance. Procurement of bad quality medicines is not only a waste of limited funds, but also results in diminished confidence in the suppliers of the medicines. This in turn causes commercial losses to companies, and affects the credibility of health programmes.

In this chapter we address the issue of medicine quality, and some of the consequences of medicine use. We begin by defining medicine quality and describing the elements of quality assurance. We then look at the problems created by the production of sub-standard medicines, and we examine the counterfeiting of medicines and its consequences and describe methods for counteracting it. We consider adverse drug reactions and systems for reporting them. Finally, we explore the problem of antibiotic resistance and the strategies that can be applied to contain it.

10.2   Quality of medicines

Pharmaceutical products can prevent or cure disease only if they are safe, effective, used rationally and of good quality. Medicines need to retain their quality from the time of manufacture to the time of consumption by the patient. They are subjected to a wide variety of hazards between manufacture and reaching the end-user, usually many months later and sometimes thousands of miles away. Smuggling and illegal importation of medicines are often rife. Substandard and counterfeit medicines are not only sold locally but are also exported.

The situation is made worse by the fact that medicines exported from industrialized countries are not regulated in the same way as those for the domestic market. At the same time the export of medicines to developing countries via free trade zones is increasing. Re-labelling of products to mask details of origin also occurs. Increasingly medicines are distributed via the internet, over which there is little control. The quality of pharmaceuticals is increasingly becoming a global concern.

Poor medicine quality has the potential to result in a number of serious consequences. These include:

· a significant *economic impact* (cost) with waste of limited resources;
· *clinical impact* (patient suffering) with lack of therapeutic effect, prolonged illness and death; and
· *health system impact* with loss of credibility in the health care system.

National medicine regulatory authorities and WHO have implemented various approaches to ensure acceptable quality of medicines. Within a national regulatory framework, registration, inspection and testing have been widely used for quality assurance in most countries. Each of these is usually in a different phase of development and implementation in a particular country, and each has had varying degrees of success. The strategies adopted have varied according to the socioeconomic and institutional infrastructures in each country, and are often limited by lack of adequate resources and technical capabilities. Given the scarcity of resources, prioritizing strategies to address quality problems becomes increasingly important for all governments and organizations involved with medicine supply.

Box 10.2: Characteristics of medicine quality

Identity:	Refers to correct active ingredient.
Potency:	Refers to the declared active ingredient; *content range*: usually 95–110. per cent as defined in pharmacopoeias allowing a margin of safety.
Purity:	To exclude contaminating substances or micro-organizms.
Uniformity:	Concerns the consistency of colour, shape, size of tablets etc that should not vary. Deviation may suggest problems with identity, purity and potency.
Bioavailability:	Is defined as the rate and extent of absorption of a medicine into the body.
Stability:	Refers to expiry, shelf life.

10.2.1 Defining medicine quality

National pharmacopoeia commissions and WHO play an important part in ensuring the quality of medicines. Detailed descriptions of medicine characteristics and of analytical techniques to verify these characteristics are laid down in national reference books called pharmacopoeias. Most manufacturing or exporting regions have their own pharmacopoeias; examples include the United States Pharmacopoeia (USP) and the British Pharmacopoeia (BP). These specify quality standards, with approved limits and other specifications, together with analytical techniques for medicines in those countries. WHO's International Pharmacopoeia is suited to the needs of developing countries in that it is based on standard chemical tests which can be carried out without sophisticated equipment.

Pharmacopoeias contain specifications for a wide variety of medicine characteristics. The most important of these characteristics are summarised in Box 10.2.

The quality of medicines reaching the patient can be affected by the manufacturing process, packaging, transportation and storage conditions, handling and other factors. These influences can be cumulative. Clinically relevant instability is rare, but poor initial quality is a more serious problem. Quality problems can include:

· Loss of potency due to poor bioavailability, expiry, or poor storage conditions;
· Too low or too high concentration due to manufacturing or compounding error and counterfeiting;
· Degradation into toxic substances has been described for tetracyclines;
· Adverse reactions;
· Contamination with bacteria or fungi of injectables, creams, syrups and eye drops.

Box 10.3: Problems with substandard drugs

Percentage breakdown of data on 325 cases of substandard drugs, including antibiotics, antimalarial and antituberculosis medicines, reported from around the world to the WHO database:

no active ingredient	60 %
incorrect amount	17 %
incorrect ingredient	16 %
other errors	7 %

Source: WHO/EDM website (2003) http://www.who.int/

The extent of the problem of substandard medicines in relation to antibiotics, antimalarial and TB medicines is illustrated in Box 10.3.

10.2.2 Quality control and quality assurance

Two important concepts need to be distinguished.

· *Quality assurance* (QA) is the whole process of assuring quality from manufacturer to end-user.
· *Quality control* (QC) involves verifying and testing the quality of the final product.

The purpose of QA is to make certain that each medicine reaching the patient is safe, effective and acceptable. QA includes technical, regulatory, operational and managerial activities spanning the entire supply chain, from medicine manufacturing to patient use.

One important determinant of both quality assurance and medicine quality is effective *medicines regulation*. However, fewer than one in six countries have well-developed medicines regulatory systems. Many developing countries have either no medicines regulatory authority at all, or one with very limited capacity that has little impact. With no means of enforcement pharmaceutical manufacturers, importers and distributors do not always comply with regulatory requirements, and the quality, safety and efficacy of both imported and locally manufactured medicines cannot be guaranteed.

There is no shortage of guidelines relating to the quality assurance of medicines. Indeed, large numbers have been developed by WHO with support from specialists in industry, national institutions and NGOs. Their main focus is on national medicine regulation, product assessment and registration, distribution of medicines, the International Pharmacopoeia, basic tests and laboratory services.

10.2.3 Elements of quality assurance

Three main functions are necessary to ensure the quality of medicines. These are:

- *registration* of medicines with the national medicines regulatory authority;
- *inspection* of manufacturing plants to ensure compliance with GMP guidelines (see below): inspection of each link in the distribution chain is also necessary; and
- *testing* of medicines (QC) based on product types and characteristics.

QA systems can be very sophisticated and costly. Where resources are available all three functions are of equal priority. In countries with scarce resources, registration is the first priority, followed by inspection and then testing. The registration of medicines has previously been described in chapter 6.

A range of approved *documents for product certification* are available. These include registration licences and Good Manufacturing Practice (GMP) certificates from medicines regulatory authorities, batch certificates from manufacturers, and the WHO certification scheme on the quality of pharmaceutical products moving in international commerce.

Under this scheme WHO put into place a uniform system of certificates issued by national regulatory authorities. The scheme has been available since 1975 as a means of exchanging information between regulatory authorities in exporting and importing countries; 139 countries currently participate.

The purpose of the WHO scheme is to confirm that the medicine has been authorized for marketing in the exporting country, and that the manufacturer is subject to regular inspections and demonstrates conformity to GMP. It also provides a mechanism for the exchange of information. As with all documents, the information contained in it is only as reliable as the authority issuing it.

An important element of quality assurance is *GMP*, determined by inspection of manufacturing sites. GMP is a system for ensuring that medicinal products are consistently produced and controlled according to quality standards. GMP covers all aspects of production and includes guidelines concerning personnel, facilities, equipment, sanitation, raw materials, manufacturing processes, labelling and packaging, quality control systems, self inspection, distribution, documents/records and complaints and adverse medicine reaction systems.

To ensure quality control throughout the distribution chain laboratory facilities are necessary to test medicines. Quality control can include physical inspection of each shipment and testing by exception: this means that samples are sent for testing when a question is raised concerning quality.

Since resources are generally limited, priorities for quality assurance have to be set. It is therefore important to focus on medicines in the following categories:

- those that have the greatest health impact;
- those that have the greatest budgetary effect;
- those that have a small therapeutic window;

- those that have bioavailability problems;
- those that have stability problems;
- sustained-release products;
- medicines from new suppliers;
- donated medicines.

QA activities concern every step in the medicine supply management cycle. They can be differentiated into technical activities, such as evaluation of product documentation and performing laboratory testing, and managerial activities, which include the selection of suppliers and performance monitoring, as well as medicine inspection procedures throughout the distribution chain.

10.2.4 Indications for quality control testing

Laboratory testing is costly in terms of human and technical resources. The actual tests performed depend on the medicinal product and the reason for testing. Quality control is particularly indicated in certain situations and should be targeted:

- to verify quality of shipment;
- for supplier selection;
- customer complaint;
- donations;
- stability and bioavailability problems;
- expired medicines;
- no or non-functional medicine regulatory authority;
- counterfeit medicines.

10.2.5 Quality assurance programmes

In order to promote good medicine quality attention needs to be paid to every step in the medicine supply management cycle. During selection, preference can be given to medicines with a long shelf life (examples include using powders for reconstitution). Bioequivalence of products should be ensured, since this is a problem with some medicines. Suppliers should be carefully selected, possibly by means of prequalification procedures (for example, the use of restricted tenders). New suppliers should be sought continuously for prequalification to maintain competition.

WHO has initiated prequalification projects for the procurement of HIV/AIDS, tuberculosis and malaria medicines. In case of open tenders, recent GMP inspection reports can be obtained from the regulatory authority, and reference checks can be carried out through agencies such as the International Dispensary Association (IDA). Care should be taken in the drawing up of contract specifications, which should include requirements in terms of pharmacopoeia reference standards, labelling and packaging.

Regulatory decisions on medicines have been compiled in the "UN Consolidated list of products whose consumption and/or sale have been banned, withdrawn, severely restricted or not approved by governments". WHO has published two updates to this list entitled "Pharmaceuticals: Restrictions in use and availability".

A comprehensive quality assurance program must include procedures covering selection and procurement of good quality medicines, verification of received medicines as well as maintaining and monitoring of medicine quality. A reporting system is essential for the follow-up of medicine quality problems. Practical recommendations include:

- *medicine selection*: select medicines based on safety and efficacy, in dosage forms with longest shelf life.
- *supplier selection*: select only acceptable and reliable suppliers based on pre-qualification, documentation, GMP, WHO certificates, informal information from other buyers or information from WHO or MRA. Insist on detailed contract specifications (e.g. labelling, pH norm, shelf life).
- *shipment inspection*: check specified quality standards at time of delivery with random physical inspection and laboratory testing.
- *assure good packaging*: organize handling and storage with an orderly system, protection (pest, heat, expiration, fire, moisture, light, theft) and record keeping.
- *repackaging and dispensing*: monitor practices.
- *storage and transport*: ascertain that conditions for these are adequate.
- *reporting system*: ensure that systems are in place for reporting product quality problems and that problems are addressed, resolved and documented.

A QA program must also include training and supervision of staff involved at all levels of the manufacturing and supply process. In addition, an effective information system is needed to follow up and document quality problems. Whilst quality assurance is founded on regulations and standards, it is the people who enforce the regulations or work to comply with the standards who make the difference between quality assurance or lack of it.

10.2.6 Follow-up of quality problems

Medicine quality problems are common, but actions to remedy and follow up these deficiencies are often not taken due to lack of a structured and institutionalized quality assurance system with standard operational procedures. Depending on the severity and potential adverse health impact of a quality problem the following activities may be recommended:

- stop distribution;
- recall batch;

· repeat testing;
· send to a reference laboratory;
· complain to the manufacturer;
· inform the DRA;
· take legal action.

10.3   Counterfeit medicines

A separate and distinct issue from that of substandard medicines arising from poor medicine quality assurance procedures at all stages in the medicine supply management cycle is the proliferation of counterfeit medicines. The counterfeiting of pharmaceuticals and the proliferation of substandard medicines constitute a serious health risk for the world population.

A counterfeit medicine is one that is deliberately and fraudulently mislabelled with respect to its identity and/or source. Trade in counterfeit medicines is a serious crime. Counterfeiting of pharmaceuticals is often undertaken by people and organizations involved in other types of crime. It is estimated that 7 per cent of medicines sold worldwide are counterfeit, with a value of more than US $ 20 billion. Counterfeiting of medicines is on the increase and it is estimated that more than 70 per cent occurs in low income countries.

10.3.1   Factors contributing to counterfeiting of medicines

Trade in counterfeit medicines is widespread and affects both developing and developed countries. It is much more prevalent in developing countries where the following conditions often occur:

· weak medicine regulatory control and enforcement;
· scarcity and/or erratic supply of basic medicines;
· demand exceeding supply;
· unregulated markets and distribution chains;
· high medicine prices;
· significant price differentials; and
· corruption and conflict of interest.

Both branded and generic medicines are subject to counterfeiting, which takes a wide variety of forms. Generally, counterfeit products may include products with the correct ingredients or with the wrong ingredients, without active ingredients, with incorrect quantities of active ingredients or with fake packaging. The majority of counterfeit cases involve tablets and capsules. Antibiotics account for almost half of the reported cases of counterfeit medicines. The type of counterfeiting, and the sorts of medicines counterfeited, as reported to WHO, are illustrated in Box 10.4.

Box 10.4: Classes of counterfeit medicines

Classes of medicines reported as counterfeit between January 2000 and December 2001

Therapeutic category	% of all counterfeit reports
Antibiotics	28 %
Antihistamines	17 %
Hormones	12 %
Steroids	10 %
Vasodilators	7 %
Drugs for erectile disorder	5 %
Antiepileptics	2 %
Others	19 %
Total	100 %

Source: WHO/EDM website (2003) http://www.who.int/

10.3.2 Consequences of counterfeit medicines

In most cases, counterfeit medicines are not equivalent in safety, efficacy and quality to their genuine counterparts. Even if they are of the correct quality or contain the correct amount of active substance, their production and distribution are not within the control of the medicines regulatory authority of the country concerned. Counterfeiting poses a health risk for patients and diminishes their confidence in medicines suppliers, which causes commercial losses to manufacturers.

Counterfeit medicines are dangerous and their use can result in death. It is a major cause of waste of public and private funds. It can not only impact on the health of individual patients, but also favours the resurgence of diseases. Vaccines may be ineffective, and anti-microbial resistance may be promoted if inadequate amounts of active ingredients are contained in counterfeit antibiotics. Examples of deaths resulting from the use of counterfeit medicines include:

- The consumption of a paracetamol syrup prepared using glycerol which was contaminated with diethylene glycol (a toxic chemical used as anti-freeze) resulted in the death of 89 children in Haiti in 1995.
- Around 2,500 people are believed to have died in Niger in 1995 after they were given a fake meningitis vaccine.

Box 10.5: WHO steps to curb counterfeiting

· National laws should regulate manufacture, trade, distribution and sale of medicines effectively, with severe penalties for manufacturing, supplying or selling counterfeit medicines.
· National medicines regulatory authorities (MRA) responsible for the registration and inspection of locally manufactured and imported medicines should be strengthened.
· They should develop standard operating procedures and guidelines for the inspection of suspected counterfeits, and should initiate widespread screening tests for the detection of counterfeits.
· Adequate training and powers of enforcement against counterfeits should be given to personnel from medicine regulatory authorities, the judiciary, customs and police.
· Partnerships should be established between health professionals, importers, industry and local authorities to combat counterfeits.
· Countries should systematically use the WHO Certificate Scheme on the Quality of Pharmaceutical Products Moving in International Commerce. Countries in the same region should work towards the harmonization of their marketing authorization procedures.
· Countries should exchange experience and expertise in areas related to quality control, medicine detection and enforcement.

Source: WHO/EDM website (2003) http://www.who.int/

Up to April 1999, 771 cases of counterfeit medicines were reported to WHO. Of these, 325 were tested for the quality of their supposed active ingredients. Only 7 per cent contained the correct amount of active ingredients. About 59 per cent contained no active ingredients, 17 per cent contained the incorrect amount of active ingredients and 16 per cent contained different active ingredients.

10.3.3   Counteracting counterfeit medicines

There is no simple solution that can be used to eliminate counterfeit medicines, nor can the problem be solved by an individual company or government. The problem has reached global proportions and requires a global approach. According to the WHO/EDM Action Plan, a number of important steps should be undertaken to curb counterfeiting of medicines. These are listed in Box 10.5.

Many countries are stepping up their offensives against this serious problem. In Russia, legislative reforms are fast-tracked, and plans are underway for a network of quality control laboratories checking strategic points in the pharmaceuticals supply channels. A new state pharmaceutical inspection unit has been established, and an amended certification system and new screening test systems have been introduced.

10.4   Adverse drug reactions

With the use of any medication comes the possibility of unintended consequences. These events, when harmful, are referred to as adverse drug reactions (ADRs). Whilst the nature of the intended benefit from using the medication is known, ADRs can include both predictable and unpredictable events. Health professionals have an important role to play in the process of continuing surveillance and monitoring of safety and efficacy of medicines that are used in clinical practice.

A medicine that has undergone all stages of preclinical and clinical trials, has been approved by a Medicines Regulatory Authority (MRA), and has successfully been marketed for a given indication cannot be considered safe and effective without reservation. Different populations, concurrent diseases, combinations of medicines, as well as rare, unintended or unexpected ADRs and toxicity can emerge after a medicine starts to be used in clinical practice. It is therefore most important that health professionals become familiar with ADRs, and are aware of their professional obligation to detect and report them, to reduce morbidity and mortality by early detection of medicine safety problems.

10.4.1   Definitions

An ADR may be defined as "any response to a medicine which is noxious, unintended and occurs at doses normally used for prophylaxis, diagnosis or therapy". ADRs are therefore unwanted or unintended effects of a medicine, including idiosyncratic effects, which occur during its proper use. They should be distinguished from accidental or deliberate excessive dosage, or medicine administration errors.

Two main types of ADRs are described.

- In *Type A reactions* the effects are directly linked to pharmacological properties of the medicine. An example is hypoglycemia induced by an antidiabetic medicine.
- In *Type B reactions* the effects are unrelated to the known pharmacology of the medicine. An example is anaphylactic shock induced by penicillin.

A *serious adverse event* is any event that is fatal, life threatening, permanently or significantly disabling, that requires or prolongs hospitalization, causes a congenital anomaly or requires intervention to prevent permanent impairment or damage.

To prove that an adverse reaction is causally related to a suspected medicine certain criteria have to be fulfilled. The effect has to end after withdrawal of the medicine and has to reappear after re-challenge with the same medicine. The probability of *causality of an ADR* is commonly described using a variety of terms including: certain, probable/likely, possible, unlikely, conditional/un-classified and un-assessable/un-classifiable.

Box 10.6: Frequency of adverse drug reactions

Very common*	≥ 1/10	(≥ 10 per cent)
Common (frequent)	≥ 1/100 and < 1/10	(≥ 1 per cent and < 10 per cent)
Uncommon (infrequent)	≥ 1/1,000 and < 1/100	(≥ 0.1 per cent and < 1 per cent)
Rare	≥ 1/10,000 and < 1,000	(≥ 0.01 per cent and < 0.1 per cent)
Very rare*	< 1/10,000	(< 0.01 per cent)

*Optional categories
Source: WHO/EDM website (2003) http://www.who.int/

10.4.2   Predisposing factors

Some factors can predispose to adverse effects. It is well known that different patients often respond differently to a given treatment regimen. In addition to the pharmacological properties of the medicine there are characteristics of the patient that predispose to ADRs. The genetic makeup of an individual patient, or concurrent diseases other than the one being treated, can also make a patient more prone to ADRs.

The very old and very young are more susceptible to ADRs. Medicines that commonly cause problems in the elderly are hypnotics, diuretics, NSAIDs, antihypertensives, psychotropics and digoxin. Medicines associated with ADRs in children include valproic acid, chloramphenicol (grey baby syndrome), antiarrhythmics (worsening of arrythmias) and aspirin (Reye's syndrome).

10.4.3   Interactions

Medicine-medicine interactions are some of the most common causes of adverse effects. When two or more medicines are administered to a patient, they may either act independently of each other or interact with each other. Interaction may increase or decrease the effects of the medicines concerned or may cause unexpected toxicity. Interactions can also occur with non-prescription medicines, social medicines such as alcohol and traditional medicines.

10.4.4   Frequency reporting

In the medical literature and in package inserts for patients, incidence and frequency of ADRs are usually listed according to international standards. It is always difficult to estimate incidence on the basis of spontaneous reports, owing to the uncertainty inherent in estimating the denominator and the degree of under-reporting.

Box 10.7: A step-wise approach to assessing possible ADRs

---

1. Ensure that the medicine ordered is the medicine received and actually taken by the patient.
2. Verify that the onset of the suspected ADR was after the medicine was taken and carefully discuss the event with patient.
3. Determine the time interval between the beginning of medicine treatment and the onset of ADRs.
4. Evaluate the suspected ADR after discontinuing or reducing the dose and monitor the patient's status. Unless otherwise indicated restart (re-challenge) the medicine and closely monitor recurrence of any ADRs.
5. Analyse alternative causes other than the suspected medicine that could cause the reaction.
6. Obtain information in the literature on any similar reports on the given reaction. Other resources may be the MRA or the manufacturer of the medicine.
7. Report any suspected ADR to the nominated person or centre.

Source: WHO website (2003) http://www.who.int/

Precise rates will inevitably be based on studies, and be limited to the more common reactions. An estimate of frequency should be provided in a standard form wherever possible.

Categories of frequency are recommended in the Guidelines for Preparing Core Clinical Safety Information on Medicines produced by a WHO Working Group in 1995. These are shown in Box 10.6.

10.4.5 Magnitude of the problem

A number of studies in the last decade have demonstrated that medicines related morbidity and mortality is a major health problem that is beginning to be recognized by health professionals and the public. It has been estimated that ADRs are the fourth to sixth largest cause for mortality in the USA. They result in the deaths of several thousands of patients each year. The percentage of hospital admissions due to ADRs in some countries is more than 10 per cent. In the UK it is estimated that 16 per cent of admissions are due to ADRs, in France it is around 13 per cent and in Norway 11.5 per cent.

In addition to suffering, prolonged and new morbidity, the economic consequences of ADRs are significant. However, there is little information in the literature about the impact of ADRs in developing or transitional countries. But it is believed that the problem is even worse as a result of the level of irrational medicine use in general. The problem is compounded by the fact that in some countries with limited resources medicine regulation is poor, medicine counterfeiting is common and independent medicine information is rare.

Box 10.8: What to report

· All even mild ADR related to new medicines.
· Unusual or serious ADR of old medicines.
· Increased frequency of certain ADR.
· ADR associated with drug-drug or drug-food interactions.
· ADR in special fields such as pregnancy lactation or drug abuse.
· ADR related to medicine withdrawal.
· ADR occurring from medicine overdose or medication error.
· Lack of efficacy of a medicine.

Source: WHO/EDM website (2003) http://www.who.int/

Despite very sophisticated and thorough clinical trials there is no guarantee that all adverse effects of a new medicine can be detected. Thalidomide was the first recognized public health disaster of a newly introduced medicine. It took decades before the ADRs of aspirin on the gastro-intestinal tract became apparent. It also took years to recognise the renal toxicity of phenacetin.

ADRs may be easily confounded with disease processes and are sometimes difficult to distinguish and single out from other causes. A step-wise approach has been suggested to assess possible ADRs. This is illustrated in Box 10.7.

10.4.6   Reporting adverse drug reactions

The monitoring of adverse drug reactions is now highly sophisticated, and a range of methods have been developed to report and analyse them. These include pharmacovigilance and post-marketing surveillance, and we describe these in chapter 12.

Many countries have established drug monitoring systems for early detection and prevention of medicine related problems. Some countries have policies and laws that oblige pharmaceutical manufacturers to report ADRs to the MRA or pharmacovigilance centres. Mandatory reporting of ADRs is of great importance not only for public health but also for the pharmaceutical industry, which has a particular interest in questions of liability due to adverse events with medicines.

But the effectiveness of both pharmacovigilance and post-marketing programmes is directly dependent on the active participation of health professionals. All healthcare providers including medical practitioners, pharmacists, dentists and nurses should report ADRs as part of their professional responsibility. Even in the absence of a clear cause for an ADR, or doubts about events, it is important to report suspicions. The pharmacovigilance centre will be able to assess and classify a suspected ADR. Reporting of ADRs can have an important influence on subsequent labelling, as well as professional advice and patient information. For example the antimicrobial levofloxacin was marketed in the USA in 1997. As a direct result of good reporting of ADRs an additional caution for cardiac patients was included on the label.

10.4.7   How and what to report

All suspected ADRs that can be considered of clinical importance should be reported. A list of the kind of reactions which need to be reported is given in Box 10.8.

Arrangements for reporting ADRs vary from country to country. Many countries have standardised forms that need to be filled with the pertinent information and sent to the relevant authority or centre. These case report forms usually include at least the following:

· patient information;
· detailed description of the adverse event and outcome;
· suspected medicine, manufacturer, indication, dosing and concurrent diseases and medicines;
· name of reporter.

10.5   Antimicrobial resistance

Communicable diseases remain a significant cause of disability, are responsible for continued high mortality, and mostly afflict the world's most vulnerable populations. Antibiotics have completely transformed our approach to infectious disease. The use of antimicrobials has led to a dramatic decline in once common and often fatal infectious diseases. Our ability to control infectious disease is, however, a recent development, and unfortunately we are already faced with a new crisis.

The development of antimicrobial resistance means that diseases which until recently were close to being controlled, contained and curable, are now difficult to treat and likely to re-emerge. Under-use and misuse of antimicrobials leads to countless deaths of people worldwide from preventable and curable conditions. With the exception of anti-retroviral medicines, research into antimicrobial therapy has slowed significantly in recent years. This is due to confidence in the effectiveness of existing antimicrobials, and a shift in emphasis to research on more profitable medicines for chronic non-communicable diseases.

10.5.1   The discovery of antibiotics

The twentieth century saw a revolution in discovery and marketing of new medicines, including antimicrobials, which transformed the treatment of infectious diseases. Bacterial and fungal infections can be treated with antibiotics: vaccines offer protection against other infections, including viral agents. Only smallpox vaccine, quinine and penicillin were discovered by chance. Ehrlich discovered the first effective treatment of syphilis with arsphenamine (Salvarsan) in 1908. Fleming made his observation of the effects of penicillin in 1928. The antibacterial actions of the sulphonamides were discovered in the late 1930s. These were rapidly followed by more antibiotics, antifungals and antiparasitics.

Together these new agents have saved millions of lives and prevented innumerable disabilities. Most were discovered between 1930 and 1970. The last 30 years have seen the development of antivirals and antiretrovirals as a further advance in the fight against infectious diseases caused by the HIV pandemic. Today about 150 antimicrobial compounds are available on the world market. However, few of these breakthroughs in infectious disease have been made accessible to all people, particularly those in developing countries.

10.5.2   The emergence of antimicrobial resistance

It was during the 1950s that scientists first became aware of the existence of a penicillin-resistant strain of *Staphylococcus aureus*: since then the emergence of resistant strains has been relentless. Resistant strains of gonorrhoea, shigella and salmonella followed. Multiple medicine-resistant tuberculosis (MDR-TB) has spread to locations around the world. Penicillin-resistant pneumococci, resistant cholera and malaria are on the rise, and formerly effective medicines are largely useless to contain these infections. In South-East Asia 98 per cent of all gonorrhoea cases are multi drug-resistant. In the industrialized world as many as 60 per cent of hospital acquired infections are caused by drug-resistant microbes and have spread in the communities.

The most serious infections concern vancomycin-resistant enterococcus (VRE) and methicillin-resistant *Staphylococcus aureus* (MRSA). Up to 70 per cent of cases of acute respiratory infections caused by *Streptococcus pneumonia* and *Haemophilus influenzae* (the primary bacteria implicated in these infections) are now resistant to first line antibiotics. A major cause of this phenomenon is the irrational treatment of viral infections of the respiratory tract with antibiotics. This is not only ineffective but also contributes to resistance. A growing number of AIDS patients show resistance to antiretrovirals. Malaria is reappearing in areas formerly considered disease free. Resistance to chloroquine, the medicine of choice, is widespread in 80 per cent of the countries where malaria is endemic, whilst resistance to newer second and third line medicines continues to grow.

10.5.3   Factors contributing to antimicrobial resistance

Resistance to antimicrobials is developed through natural selection, with more susceptible organisms succumbing and more resistant ones surviving. Some of the factors that contribute to antimicrobial resistance are listed in Box 10.9.

Box 10.9: Factors contributing to antimicrobial resistance

· Environment and society.
· Poverty and access.
· Misdiagnosis and inappropriate prescribing.
· Counterfeit medicines.
· Advertising and patient demand.
· Lack of education.
· Prescribing in hospitals.
· Livestock and food.
· Globalisation and tourism.
· Socio-cultural attitudes and behaviour.

Source: WHO/EDM website (2003) http://www.who.int/

A major problem in the development of resistance is poverty, leading to limited access to medicines and treatment. Even if antimicrobials are available, patients often cannot afford to pay for a full course of antibiotics, and so buy just enough for a few days' treatment, or resort to substandard or counterfeit medicines. About 5 per cent of all antibiotics sold worldwide are counterfeit or substandard. Resistance emerges whenever antibiotics are used at doses lower than indicated. A majority of counterfeit medicines are to be found in developing countries.

Misdiagnosis by poorly trained health staff, often with inadequate diagnostic facilities to test for culture and sensitivity, leads to inappropriate prescribing. In addition, patient pressure to prescribe antibiotics and fears of liability may add to unnecessary prescribing. These factors frequently lead to the prescribing of expensive, broad-spectrum antibiotics, even when cheaper narrow spectrum antibiotics might suffice. In many countries free availability of antibiotics in medicine stores leads to unqualified self-diagnosis and overuse of antibiotics. In a study in Vietnam in 1997, researchers found that more than 70 per cent of patients were prescribed inadequate amounts of antibiotics for serious infections, and 25 per cent were given unnecessary antibiotics. The same is true for North America where medical practitioners over-prescribe antibiotics by as much as 50 per cent.

The lack of education of health workers is another factor contributing to irrational antibiotic use. Worldwide, antimicrobial pharmacology and therapy is inadequately covered in medical schools. In an analysis of 10 studies undertaken at teaching hospitals worldwide, it was found that between 40 and 91 per cent of antibiotics were prescribed inappropriately. The survey also revealed that health workers often neglect basic hygiene practices, and that this has serious consequences both for the hospitals and the community. Hospitals are a breeding ground for antibiotic-resistant bacteria. Costs of hospital-acquired infections have been estimated at US $ 10 billion per year in the USA and US $ 450 million per year in Mexico.

Box 10.10: Action plan to control antibacterial resistance proposed by WHO

· Adoption of *international policies and strategies* to prevent, treat and control infectious disease. This includes vaccination, treatment guidelines such as the Integrated Management of Childhood Illness (IMCI) or the DOTS strategy for tuberculosis and surveillance systems.
· *Education* of health staff and the public on rational use of medicines.
· Contain resistance in the *hospitals*.
· Reduce the use of antimicrobials in *livestock and food*.
· Increase *research* for new medicines and vaccines.
· Partnerships to increase *access* particularly for poor populations.

Source: WHO/EDM website (2003) http://www.who.int/

About 50 per cent of all antibiotics produced are used to treat animals, for growth promotion in livestock, prophylaxis and for decontamination of various foodstuffs. International travel, immigration and tourism also play a role in the development of resistance. In certain infections such as tuberculosis the problem lies in the long treatment periods required, which often lead to poor compliance, limited access and the use of poor quality medicines.

10.5.4   The consequences of antimicrobial resistance

Antimicrobial resistance has become a serious and global public health concern with economic, social and political implications across borders. The consequences of antimicrobial resistance are multiple, and are as follows:

· mortality: resistant infections are more often fatal;
· morbidity: prolonged illness, greater chance for resistant organisms to spread to other people;
· cost: increased cost of care, newer, more expensive medicines;
· limited solutions: few new medicines on the horizon.

Ineffective medicines have to be replaced by newer medicines up to a hundred times more expensive: a single six-month treatment using standard anti-tubercular medicines costs about US $ 20, whereas treatment of multiple drug-resistant tuberculosis can cost more than US $ 2,000. This in turn will lead to under-use, courses of treatment will not being completed, and yet further resistance will result. Ineffective treatment causes increased mortality. Prolonged illness will increase the risks of contamination, and therapeutic alternatives will become limited and more expensive.

Box 10.11: Key issues concerning antimicrobial resistance

· Antibiotics enable huge advances in medicine.
· Poor antibiotic use means survival of resistant bacteria.
· Resistant bacteria accumulate and spread.
· Resistance increases clinical complications, lengthens hospital stay and adds cost.
· Development of new antibiotics is slow, expensive and cannot be guaranteed.
· With more resistance and fewer new antimicrobial agents, modern medicine will undergo significant setbacks.

Source: WHO/EDM website (2003) http://www.who.int/

10.5.5   Strategies to contain antimicrobial resistance

Unprecedented availability of antimicrobials and their impact on quality of life has led to complacency and misuse of resources. A proper understanding of the factors that influence antibiotic use should be seen as a precondition for the development of effective policies and programmes to address inappropriate antibiotic use. If interventions into antibiotic use are to be effective, future research needs to explore in more depth the socio-cultural rationality of antibiotic usage.

Large investments of effort, funding, time, commitment and cooperation are needed to contain antimicrobial resistance. Approaches include training of health staff, better treatment modalities, community education, immunisation programmes, improved hygiene, nutrition, and enhanced vector control. Importantly there needs to be concerted action by governments, health professionals, the community, the pharmaceutical industry and NGOs. Box 10.10 summarises an action plan to control antibacterial resistance proposed by WHO.

Resistance can be effectively approached and treated by rational medicine use, involving the correct medicine administered by the best route, in the right dose and at the optimal intervals for the appropriate period after an accurate diagnosis.

The following practical advice can be given to general practitioners and prescribers:

· *Don't* prescribe antibiotics for simple cough and colds.
· *Don't* prescribe antibiotics for viral sore throats.
· *Limit* antibiotic prescription for uncomplicated cystitis to three days in otherwise fit young women.
· *Limit* prescribing antibiotics over the phone to exceptional cases.

10.5.6 Key issues in antimicrobial resistance

Antimicrobial resistance is not a new or surprising phenomenon. All microorganisms have the ability to evolve various ways of protecting themselves from attack. However, over the last decade antimicrobial resistance has increased and the pace of development for new antimicrobials has decreased. Both ready availability and use as well as inadequate access and under-treatment with antibiotics can promote resistance. Key issues concerning antimicrobial resistance are summarised in Box 10.11.

The main priority should be to prevent infection in the first place. After this, containment of drug resistance should be the aim. Since antimicrobials drive resistance, the focus of any containment strategy needs to be to minimise any unnecessary, inappropriate or irrational use of antimicrobial medicines. All groups of people involved with antimicrobials need to be engaged: patients, prescribers and dispensers, hospital managers, agriculture, governments, the pharmaceutical industry, international agencies, NGOs and professional organizations.

10.6   Conclusion

This chapter has considered the complex and serious issues around the quality of medicines. It is not enough for populations to have access to medicines and to have strategies to ensure their rational use. These efforts will be wasted if the quality of the medicines available is sub-standard. As we have seen, quality can be compromised at every stage of the medicine supply management cycle: a wide range of strategies have been developed to minimise it. The problem of counterfeit medicines remains one of the major challenges facing those involved in the management of pharmaceuticals in international health.

We have also considered in this chapter the problems associated with adverse drug reactions and the growing crisis of antibiotic resistance. Many of these problems are common to both developed and developing countries. However, the problems associated with medicine use are presented in far greater focus in low income countries, which do not have the resources to deal with them that high income countries have. Nevertheless, as we have seen in this chapter, there are many things that low income countries can do to reduce the extent of these problems, and to minimise their consequences when they do occur.

Further reading

Brewer, T. et al. (1999) "Postmarketing surveillance and adverse drug reactions", *Journal of the American Medical Association* 281: 9.

Kanjanarat, P. et al. (2003) "Nature of preventable adverse drug events in hospitals: a literature review". *American Journal of Health-System Pharmacy* 60 (14): 1750.

Lazarou, J. et al. (1998) "Incidence of ADRs in hospitalized patients: a meta-analysis of prospective studies". *Journal of the American Medical Association* 279 (15): 1000–1005.

Moore, N. et al. (1998) "Frequency and cost of serious adverse drug reactions in a department of general medicine". *British Journal of Clinical Pharmacology* 45 (3): 301–308.

Quick, J.D., Rankin, J.R., Laing, R.O., O'Connor, R.W., Hogerzeil, H.V., Dukes, M.N.G. and Garnett, A. (eds.) (1997) *Managing Drug Supply: The Selection, Procurement, Distribution and Use of Pharmaceuticals.* Second edition. West Hartford, CT, USA: Kumarian Press.

Radyowijati, A. and Haak, H. (2003) "Improving antibiotic use in low-income countries: an overview of evidence on determinants", *Social Science and Medicine* 57 (4): 733–744.

WHO (1975) "Certification Scheme on the Quality of Pharmaceutical Products Moving in International Commerce". In: *Twenty-eighth World Health Assembly, Geneva, 13–30 May 1975. Part 1: Resolutions and decisions, annexes. Official Records of the World Health Organization* 226: 94–95. Geneva: World Health Organization.

WHO (1997) *Quality Assurance of Pharmaceuticals: A Compendium of Guidelines and Related Materials.* Geneva: World Health Organization.

WHO (2000) *Overcoming Antimicrobial Resistance.* WHO/CDS/2000.2. Geneva: World Health Organization.

WHO (2001) *Global Strategy for Containment of Antimicrobial Resistance.* WHO/CDS/CSR/DRS/2001.2. Geneva: World Health Organization.

WHO (2002) *Safety of Medicines: A Guide to Detecting and Reporting Adverse Drug Reactions.* WHO/EDM/QSM/2002.2. Geneva: World Health Organization.

Chapter 11
Managing Medicines Information

Reinhard Huss

Box 11.1: Learning objectives for chapter 11

By the end of this chapter you should be able to:

· List the principal users of medicines information.
· Apply key criteria for critical medicine information management.
· List key sources of medicines information for health professionals.
· Describe traditional and internet sources of medicine information.
· List the criteria to be used for critically reviewing the quality of medicines information. contained in publications and other information.
· List the questions that should be asked when evaluating information about new medicines.
· Describe the ATC medicines classification systems and its use.
· Define a Defined Daily Dose.

11.1   Introduction

This chapter is concerned with information about medicines, and its management. Access to adequate information about medicines is a basic precondition necessary to ensure their appropriate use, affordability, accessibility and availability, and hence to provide health care services of good quality. Both health professionals and the public have an interest in objective information about medicines as an essential health technology.

Although information alone cannot guarantee the provision of good health care services, it is an important source of power for its owner. This gives key stakeholders the ability to influence other people, and hence the institutions and organizations of the health care system. Comprehensive and comprehensible information about medicines is therefore a prerequisite if we wish to empower health professionals and citizens to choose and agree on the best treatment at the personal level, and on the optimal health care service organization at the social level.

In this chapter we consider both the users and the providers of information about medicines. We describe the principal sources of such information, review factors influencing its quality, review methods for assessing information, and describe classification systems.

11.2   The users of medicines information

We can differentiate between three principal users of medicines information. These are:
· health professionals (those providing health care),
· patients (those receiving health care), and
· the public.

The first group requires detailed scientific information about medicines, ranging from their indications, dosage and precautions to the costs of treatments. This information needs to be relevant to their professional practice, and be adapted to their training.

The second group needs information about the use of prescribed medicines and self-medication. This needs to be adapted to the educational level of the patient. The third group of public citizens requires comprehensive and comprehensible information about medicines in order to participate in the decision-making process within the health care system

11.3   Sources of information

Information about medicines can be obtained from primary, secondary, and tertiary sources. The principal forms of each of these types are illustrated in Box 11.2.

11.3.1   Interpreting different sources of medicine information

Those planning to use information about medicines need to consider carefully the source of the information provided. It is helpful here to differentiate between the positivist and phenomenological scientific perspectives: the former emphasizes positive facts and the existence of a single objective reality, whilst the latter stresses that reality is socially constructed and is dependent on the interaction of individuals.

Whilst a positivistic perspective rejects newspaper articles or television programmes as subjective and unscientific information about medicines, a phenomenological perspective views all documents as social construction. No document can be seen as the complete and accurate representation of reality, but all documents will be valuable sources for certain information.

The critical reader therefore needs to not only assess the available scientific literature but also reflect on why certain information may either be scarce or not available in health science publications. What decision-making processes might prevent research on particular health issues? To what extent and why do these issues concern:

Box 11.2: Sources of medicines information

Primary sources

Scientific journals on medicine-related subjects such as clinical trials, case reports, pharmacological and pharmaceutical research reports, unpublished scientific research reports, consultancy reports, official health statistics on the use of medicines, health service indicators linked to medicine, information on health care financing linked to medicine, reports in journals and newspapers about medicines, radio and television reports about medicines.

Secondary sources

Review articles in scientific journals, meta-analyses e.g. Cochrane collaboration, indexes e.g. Index Medicus, Medline, abstracts e.g. International Pharmaceutical Abstracts, consensus documents, annual health service reports on medicines, medicine reviews for citizens in journals on radio and television.

Tertiary sources

Formulary manual, standard treatment manual, essential and positive medicine lists, textbooks, general reference books, e.g. pharmacopoeia, medicine bulletins and newsletters, medicine compendia for professionals or the public.

- diseases that mainly affect poor people?
- how treatment with medicines can be avoided or reduced? and
- how traditional medicine can be used more effectively?

Readers therefore need to know which type of information they are looking for, and to question the information source with a healthy scepticism. Box 11.3 lists some key questions that need to be asked about information sources before attaching any degree of reliability to them.

The social context and the process of information construction should be critically assessed, especially when the information is provided by stakeholders with vested interests. For example, some questions which may need to be asked about the information provided by pharmaceutical manufacturers are listed in Box 11.4.

11.3.2   Sources for health professionals

Three sources of information about medicines may be of particular interest to health professionals, as they may reduce the amount of time and work needed to keep up to date with medicines information. These are consensus documents, meta-analyses, and medicine bulletins or newsletters.

- *Consensus documents* are useful when there is a lack of definitive evidence about interventions in clinical medicine or health policy. However, there is

---

Box 11.3: Questions to ask about information sources

Primary sources
· What is the reputation of the scientific journal? Is it a peer-reviewed journal?
· What is the reputation of the newspaper, journal, TV or radio station? Are they adherent to the standards?
· What is the author(s)' affiliation?
· If it is a scientific article, is it designed as a clinical study or a case report?
· Regarding a study: What about the sample size? How were participants selected? Is it a control, prospective or retrospective, and/or blinded study? Are the references representative?
· If you are dealing with another article or report: Is it a sensational, informational or educational presentation? Who funded the study, articles or report? Are there any conflicts of interest?
· General questions regarding any article or report: Are the conclusions reasonable?
· Do the official health statistics and reports have an internal consistency or are they missing information?
· Regarding a consultancy report: Who is funding the report? Is there an internal and external consistency?

Secondary sources: abstracting, bibliographic or indexing services
· What journals are covered, and are they inclusive and comprehensive?
· Lag time: Original publication and inclusion in service?
· Are they user-friendly, and include key words, subject headings etc.?
· Are abstracts available, and is the source accurate?

Secondary and tertiary sources: written by one or several individuals
· What are the qualification/s of the author/s?
· What is the reputation of the publisher?
· Who has financed the publication or conference?
· Is it a peer-reviewed publication?
· What are the dates of development and publication?
· Are the references included or accessible?

---

some debate about the reliability and validity of the applied methods, and the results should always be checked against observation.

· A *meta-analysis* reviews several studies on the same topic in order to pool the data and analyse it as one large study. Whilst it is a very useful approach it also has a number of limitations and problems.

· *Medicine bulletins and newsletters* provide summaries and up to date information about therapeutic progress. Examples include the Drugs and Therapeutics Bulletin (United Kingdom), Medical Letter (USA), Prescrire International (France) and Arznei-Telegramm (Germany). Such bulletins are independent of pharmaceutical industry or government interests and they provide objective and reliable information.

Box 11.4: Manufacturers' information

The following questions should be asked when evaluating the quality of medicines information provided by pharmaceutical manufacturers:

· Is the information given approved by the country's medicine regulatory authority?
· Are the references provided to peer-reviewed medical literature?
· Do they include unsolicited or requested information?
· Do they contain negative references to alternatives?
· Is it a sensational or very positive presentation?
· Do they include references to adverse drug reactions?
· Is a cost comparison included?
· Does the study correspond with current practice and standards?
· Is the information dated and updated?

Although on the face of it each of these sources offers the prospect of valid and unbiased information each needs to be interpreted with caution. Questions that should be asked of these three sources of medicine information are listed in Box 11.5.

11.3.3 The internet as source of medicine information

The internet is a relatively new medium from which a vast amount of information about pharmaceuticals can now be obtained, through mechanisms such as emails, the world wide web and usenet news (computers linked through the internet to exchange information on a particular topic). This medium has both great strengths and great weaknesses.

The *strengths* of the internet as a source of medicines information are:
· the accessibility of the system, and
· the enormous quantity of the available information.

The *weaknesses* of the internet as a source of medicines information are:
· its very diverse quality,
· problems of connectivity, and
· costs of accessibility.

The last two are particular issues in low and middle income countries. They can be major factors in preventing access to the available information. Nevertheless, access to information about medicines through the internet is better than it ever was through traditional paper-based sources.

There are at present only voluntary arrangements for the approval of quality and the accreditation of providers of medical information available on the internet. Any

---

Box 11.5: Consensus documents, meta-analysis and medicine bulletins

---

Consensus documents:
· How were individual experts selected?
· Is a definition of consensus provided?
· Is a methodology of consensus provided?
· Are the references provided and accessible?
· Is the consensus open to public review?
· Is the data based on all available evidence in peer-reviewed journals?
· Is the date of publication and frequency of up-dating provided?
· What is the reputation of the publisher e.g. WHO?

Meta-analysis:
· Have the aggregated studies been selected critically and cautiously?
· Certain problems such as the comparability of the data and inherent sample biases cannot be completely overcome by statistical analysis!
· As different statistical methods may have been used in the pooled studies, complicated and controversial statistical manoeuvres may be applied to bring the studies together.
· Poorly conducted studies may over-influence the analysis, if reviewers did not identify the problem with their inclusion criteria.

Medicine bulletins and newsletters:
· Understanding prescribing behaviour
· Decision and action orientation
· Selected key messages
· Attractive appearance
· Brief and simple presentation
· Use of respected references
· Independence from commercial influence and affiliation with credible organisation
· Relevance of discussed issues

---

such information therefore needs to be treated with great caution. We consider the role of the internet further in chapter 13.

11.3.4   Open access publishing

Open access publishing has been a recent important development for the democratization of medical information through the world wide web. The Public Library of Science (PLoS) in particular aims to promote the value and feasibility of such open access publications. The model involves the transfer of costs from the reader ("reader pays") to the author ("author pays"). For example, the PLoS charges the author of a paper US $ 1,500 per article, representing the actual costs of production. It is then made available free, so that everybody with a computer and internet connection can have access to it.

This is a very cost-effective type of information dissemination, because the estimated average price for circulating a scientific article as hard copy is US $ 4,500. Thus for one third of this cost research articles can be made available to all instead of to a dwindling group of subscribers. This approach requires a change of funding procedures, where agencies or organizations involved in research are willing and able to finance open access publishing by including it in the value of grants awarded.

11.3.5   Medicine information centres

A medicine and toxicology information centre may be the best solution to pool the knowledge, skills and resources required to run an adequate national medicine information service. Such a service can not only provide information about essential medicines and their rational use but also deliver updates on pharmaco-therapeutic literature and pricing of medicines. Safety alerts can be circulated about ADRs and continuing training may be offered to prescribers and dispensers.

It is important that the staff of such a centre has sufficient knowledge about how to access national and international information such as the Cochrane Collaboration and important journals which may be free or at a reduced charge for users from low income countries. The team should be competent in the analysis, summary and attractive presentation of information. A capacity for training and research is an asset for such a centre. The centre has to be funded adequately and the staff members need to be committed to their ethical responsibility towards society in order to avoid the influence of any special interest group.

11.4   The quality of medicines information

The quality of information about medicines, whether provided on the internet or by other means, is not an independent characteristic. It is related to the specific context and the decision-making process for which it is needed.

Seven criteria can be used to assess the appropriateness and quality of information about medicines. These are clinical relevance, social relevance, currency, user-specificity, independence, objectivity and absence of bias. We will now consider each of these in turn.

11.4.1   Clinical relevance

Clinical relevance refers to the improved outcome of health problems or specific diseases of patients. Effectiveness of medicines has to be evaluated based on changes in the course of a disease or a health problem, and not because of some improved arbitrary biochemical or physiological parameters in patients. A related theme is the increasing "medicalisation" of healthy people, in which minor and self-limiting ailments are turned into serious diseases requiring pharmaceutical treatment. This approach has sometimes been called "disease-mongering".

11.4.2   Social and cultural sensitivity, currency and independence

Information about medicines should be socially and culturally sensitive. Social relevance indicates that the information is appropriate for a certain social context. Any information about medicines should be up to date and include the most recent available evidence. It needs to be adapted to its specific type of user, whether this be a prescriber, dispenser, patient, politician or citizen.

Ideally information should be provided independently of any conflicting economic or other interest, although this is not always possible. Conflicts of interest should be stated, as they now are in a number of medical journals, so that at least these interests are open and transparent to the reader.

11.4.3   Objectivity

Scientific information strives for objectivity. This represents an ideal, and reality in practice may be rather different. Any production, publication and dissemination of information is determined by the values and priorities of all those involved. The sociologist Ann Bowling states that scientists cannot divorce themselves from the cultural, social and political context of their work. She recommends that scientists make their assumptions about their world explicit and strive to conduct their research as rigorously and objectively as possible.

Whether such a recommendation is entirely feasible, and whether it will help close the gap between the information needed and that provided, is rather doubtful. Very often scientists themselves do not have the final say or they may depend on funds from pharmaceutical companies. For instance, pharmaceutical research is neglecting infectious diseases that cause high mortality and morbidity among poor people in low income countries. Moreover, the production of information about how people can remain healthy and avoid the use of medicines does not appear to be a priority in most countries. Antonovsky has described this as the "salutogenic approach", in which research looks for factors in the social system, the physical environment and the individual which promote health, and which explain why certain people manage to remain healthy in the most difficult circumstances.

11.4.4   Bias

Bias is defined as a systematic distortion or deviation of information in one direction. Sackett has reported thirty five different types of bias that may affect the results of a study. In the context of information about medicines we should be particularly aware of problems such as publication bias, where certain information may not find a publisher. Equally published information may be distorted or incomplete due to particular interests of authors, sponsors, or publishers. Certain interests of an author may not be revealed in a document.

Probably the commonest form of bias is the absence, rather than the distortion, of scientific information. However, information can also be biased due to a fault in the design of a study.

11.5   The assessment of information

What skills does a user of medicines information need to develop in order to analyze documents in a critical way and to avoid manipulation? As we have seen, there are many hazards to be wary of in assessing information about medicines.

We have already noted (section 11.3.1) that care has to be taken in the interpretation of medicines information according to its source. The critical reader needs to reflect on why certain information may or may not be available, or may be very limited. The providers of information about medicines may have a vested interest in not being completely objective. Research into certain health issues may not be funded, particularly those that mainly concern diseases that affect poor people. Likewise, there may be resistance to the publication of studies that indicate how treatment with medicines can be avoided or reduced, or how traditional medicine can be used more effectively.

11.5.1   Missing results

Another problem is the delay or non-publication of certain scientific results that may be important for the critical evaluation of medical treatment. In 1997 a survey amongst academic life scientists in the USA tried to determine the extent and the reasons for delaying or withholding research results. A considerable number of scientists reported cases where publication had been delayed, or where specific data had not been published, because of particular commercial interests.

Four years later, in 2001, the editors of twelve international peer-reviewed medical journals published a commentary in the *Lancet* that clinical trials should be a tool for decision-making and not for marketing purposes of medicines. They stressed the need for responsibility and accountability amongst researchers and authors, and for transparency of sponsors in clinical trials. This warning was later confirmed by the case study presented in Box 11.6, where commercial interests were given priority over public interests.

11.5.2   Information about new medicines

A major concern of every conscientious health professional is the possibility of missing out on important information about therapeutic progress. Health providers in

---

Box 11.6: Case study

Manipulation of data in favour of COX-2-inhibitors in CLASS and VIGOR

Following the publication of two large studies VIGOR and CLASS of the selective COX-2 inhibitors rofecoxib (Vioxx) and celecoxib (Celebrex), doubts remained regarding safety advantages of these medicines. Now it has become clear that in both studies essential risk data where withheld from publication.

Cardiotoxicity: The published data of the VIGOR study reflect the cardiovascular risk potential of rofecoxib as compared to the non-steroidal antirheumatic medicine naproxen in an insufficient and distorted manner...
However, a safety update submitted to the American Food and Drug Administration (FDA) by the sponsor on 13 October 2000 – six weeks before the publication of the study – clearly states that severe thrombotic cardiovascular events increased significantly...

Gastrointestinal tolerance: By now, the authors of the CLASS study are accused of an attempt to mislead the public. Only results from the first six months of the study were submitted for publication to the Journal of the American Medical Association (JAMA). The author of the accompanying editorial also received only the half year data, although the study, which ran over one year, had already been completed at the time of publication...

Summary: Parallels are striking: Both studies publicly reported only the more advantageous results, while the complete data pool showing less favourable results was reported to the FDA only. In our view, this strategy was used to provide the manufacturers with a timely advantage in order to establish their products on the market. The obvious commercial interests reflected in VIGOR and CLASS by tampering with scientific results and disregarding safety concerns for patients, undermine confidence in the seriousness and scientific quality of all study data presented to the public.

Source: Arznei-telegramm (2001) "Manipulation of data in favour of COX-2-inhibitors in CLASS (Celecoxib Long-Term Arthritis Safety Study) and VIGOR (Vioxx Gastrointestinal Outcomes Research)" 32: 87–88.

---

low and middle income countries usually have often only very limited access to critical therapeutic literature. However, real therapeutic progress, such as oral rehydration salts, is rare and normally does not need advertising in order to become widely known.

From a public health perspective, new medicines should represent a real therapeutic advance over existing medicines. According to the International Society of Drug Bulletins (ISDB)'s Declaration on Therapeutic Advances a new medicine should provide improved efficacy, fewer adverse effects, or greater convenience to patients compared with existing medicines.

Apart from the numerous new medicines which do not represent genuine innovations, there is also an imbalance in the prioritisation of diseases. Pharmaceutical

---

Box 11.7: Questions to ask about new medicines

· Has the new therapeutic principle been thoroughly tested for at least 1–2 years?
· Have only hard clinical end points been used in clinical studies instead of surrogate criteria for the assessment of efficacy?
· Was the follow-up sufficient? Long-term treatment regimes require long-term studies.
· Is the available medicine information free of hidden commercial influences?
· Can any manipulation of data be prosecuted and punished in the country?
· Are experts or opinion leaders obliged to reveal their links to the pharmaceutical industry?
· Is an effective post-licensing pharmacovigilance system in place to reveal late and rare potential medicine risks?
· Have further operational studies been organized to demonstrate the effectiveness of this medicine in real life situations?

---

companies have identified the so-called lifestyle medicines as a priority for their research and development programmes. There is an increasing trend to redefine healthy life stages such as menopause as medical events which need to be treated with medicines, in order to create new market opportunities.

Unfortunately health professionals are now inundated with information about the launch not only of "lifestyle medicines" but also by so-called pseudo-innovations or "me too" pharmaceuticals. These offer little real improvement over existing treatments in terms of efficacy, safety or cost-effectiveness. Such products allow for higher returns on investment and bigger market shares by pharmaceutical companies. This commercial strategy is incompatible with public health objectives, as it undermines efforts to assist health professionals to be well-informed about medicine therapy, and it reduces the financial resources available for the development of real innovations.

11.5.3   Assessment of information about new medicines

In addition to the questions indicated earlier in this chapter about medicines information, further questions should be asked when a new medicine becomes available in a health care system. These are listed in Box 11.7.

11.6   Classification systems

A number of systems have been devised to classify medicines. Medicine classification systems can be divided into two main groups according to their purpose: therapeutic classification systems and medicine utilisation research.

Box 11.8: Anatomical Therapeutic Chemical Groups-First Level

A	Alimentary tract and metabolism
B	Blood and blood forming organs
C	Cardiovascular system
D	Dermatologicals
G	Genitourinary system and sex hormones
H	Systemic hormonal preparations, excl. sex hormones and insulins
J	Anti-infectives for systemic use
L	Antineoplastic and immunomodulating agents
M	Musculo-skeletal system
N	Nervous system
P	Antiparasitic products, insecticides and repellents
R	Respiratory system
S	Sensory organs
V	Various

11.6.1 Therapeutic classification systems

Therapeutic classification systems are used in formularies to structure the information for health care providers on medicines. There are a number of such systems in use.

- The *British National Formulary* (BNF) provides a structure for information on treatment guidelines as well as medicines. It is published twice-yearly, and has fifteen main therapeutic categories.
- The first edition of the *WHO Model Formulary* has twenty-seven therapeutic categories, which correspond with the structure of the *Twelfth WHO Model List of Essential Medicines*.

The main categories of the WHO Model Formulary are divided into a variable number of therapeutic sub-categories. Some medicines may appear in several therapeutic categories or sub-categories in order to make the formulary more user-friendly. For instance, acetylsalicylic acid is mentioned several times in the WHO Formulary according to the therapeutic indication of the product for rheumatic disease, pyrexia, pain, migraine and inhibition of platelet aggregation.

11.6.2 ATC classification system

The Anatomical Therapeutic Chemical (ATC) classification system has been developed as a tool for medicine utilisation research by the WHO Collaborating Centre for Medicine Statistics Methodology, in Oslo, Norway.

Box 11.9: ATC classification of amoxicillin and paracetamol

	Amoxicillin		Paracetamol
J	Antiinfectives for systemic use	N	Nervous system
J01	Antibacterials for systemic use	N02	Analgesics
J01C	Beta-lactam antibacterials, penicillins	N02B	Other analgesics and anti-pyretics
J01CA	Penicillins with extended spectrum	N02BE	Anilides
J01CA04	Amoxicillin, DDD: 1, Unit: g, Admin. route: O, P Notes: None	N02BE01	Paracetamol, DDD: 3 Unit: g Admin. route: O, R Notes: None

Abbreviations: DDD, Defined Daily Dose; Admin. route, Administration route; O, oral; P, parenteral; R, rectal

---

In this classification system medicines are categorised by means of five hierarchical levels. These are as follows:

- The *First Level* involves allocation to one of fourteen main *anatomical* groups, according to the organ or system on which it acts. These are listed in Box 11.8.
- The *Second Level* consists of pharmacological or *therapeutic* sub-groups:
- The *Third and Fourth Levels* are chemical/pharmacological/therapeutic sub-groups, and
- The *Fifth Level* is the *chemical* substance.

11.6.3   Defined Daily Dose

Box 11.9 illustrates the classification of two medicines, amoxicillin and paracetamol, in the ATC system. These examples also indicate the *Defined Daily Dose* (DDD) for both medicines. This is another important component of the system. The definition of the DDD is the assumed average maintenance dose per day for a medicine used for its main indication in adults. It is important to understand, however, that the DDD is only a statistical unit of measurement for drug utilisation studies and not a therapeutic recommendation for individual patients.

Since 1996 the WHO has recommended the ATC/DDD system as an international standard for drug utilisation studies, and the WHO Collaborating Centre for Medicine Statistics Methodology, established in Oslo in 1982, is responsible for its regular updating. The system cannot be comprehensive for all medicines, as they are only added on request of users of the classification.

The nomenclature of these medicines is normally based on the international non-proprietary names (INN) that are assigned through WHO. For certain medicines such as topical preparations, vaccines or anaesthetics DDDs are not established. Normally the ATC system allocates only one code for each medicinal product, even if it is used for more than one indication. However, if separate pharmaceutical formulations are used for different indications, more than one ATC code is allocated.

Although the ATC/DDD system is reviewed regularly, changes are kept to a minimum in the interest of long-term studies on medicine utilisation. The system is therefore recommended for medicine utilisation statistics, for reporting medicine poisoning and adverse reactions, but it is less useful for the therapeutic classification of medicines.

11.7   Conclusion

This chapter has been concerned with information about medicines, particularly in relation to low and middle income countries. We have considered the main sources of such information, and identified the main users of it. But we have emphasized the need to treat information about medicines with caution, by noting the interests that lie behind it, and the possible biases that may be present.

We have identified the criteria against which such information should be judged, and we have provided a number of tools that we hope will help readers ask the questions that need to be asked about particular types of information. Finally, we have described the main systems by which medicines have been classified.

The focus of this chapter has been on the availability of information about medicines *before* they are prescribed, dispensed and administered. But in order to ensure the rational use of medicines we also need to monitor the actual use of medicines *after* their prescription and dispensing. To gather this type of information we need to investigate the use of medicines, and that is the focus of our next chapter.

Further reading

Antonovsky, A. (1987) *Unravelling the Mystery of Health*. San Francisco: Jossey Bass Publishers.

Arznei-telegramm (2001) "Restricted list for pseudo-innovative medicines improves therapeutic quality and reduces expenditures". 32: 77–79. "Lesson to be learned from the cerivastatin (Lipobay) affair". 32: 88–89.

Blumenthal, D. et al. (1997) "Withholding research results in academic life science: evidence from a national survey of faculty". *Journal of the American Medical Association* 277: 1224–1228.

Davidoff, F. et al. (2001) "Sponsorship, authorship, and accountability". *Lancet* 358: 854–856.

Delamothe, T. and Smith, R. (2004) "Open access publishing takes off". *British Medical Journal* 328: 1–3.

Kiley, R. (1999) *Medical Information on the Internet: A Guide for Health Professionals*. Second edition. Edinburgh: Churchill Livingstone.

Moynihan, R., Heath, I. and Henry, D. (2002) "Selling sickness: the pharmaceutical industry and disease mongering". *British Medical Journal* 324: 886–891.

Sackett, D.L. (1979) "Bias in analytic research". *Journal of Chronic Diseases* 32: 51–63.

Schwabe, U. and Paffrath, D. (2001) *Arzneiverordnungsreport 2000*. Berlin: Springer Verlag.

WHO (2000) *World Health Report 2000*. Geneva: World Health Organization.

WHO (2000) *Global Comparative Pharmaceutical Expenditures, with Related Reference Information*. WHO/EDM/PAR/2000.2. Geneva: World Health Organization.

WHO (2002) *About the ATC/DDD System*. Oslo: WHO Collaborating Centre for Medicine Statistics Methodology (1). http://www.whocc.no/atcddd/

WHO (2002) *Guidelines for ATC Classification and DDD Assignment*. Oslo: Collaborating Centre for Medicine Statistics Methodology (2).

Chapter 12
Investigating the Use of Medicines

Reinhard Huss

Box 12.1: Learning objectives for chapter 12

By the end of this chapter you should be able to:

· List the main objectives of medicines utilization studies.
· Describe medicines utilization sudies that combine qualitative and quantitative data.
· List the WHO medicines use indicators.
· Define pharmacoeconomics.
· Describe the main methods of pharmacoeconomics.
· Define pharmacovigilance.
· Describe the main methods of pharmacovigilance.
· Define pharmacoepidemiology.
· Describe the main uses of pharmacoepidemiology.
· Explain the difference between health systems and health care systems research.
· List important regulatory issues explored in regulatory science.

12.1 Introduction

The last chapter dealt with information about the actions and uses of medicines and the interpretation of that information. It described the principal sources of information about medicines and their principal users. Once medicines are in use, we need information about how they are being used and who is using them. We need to know the economic costs involved, and the dangers involved in their widespread use; and we need to know what impact they are having on the health of the population.

To help answer these questions a range of methodological approaches have been developed, and the aim of this chapter is to introduce the methods involved. The chapter describes five of them: drug utilization studies; pharmacoeconomics; pharmacovigilance; pharmacoepidemiology; and health systems and policy studies.

12.2   Drug utilization studies

Drug utilization has been defined in general terms as "the prescribing, dispensing, administering and ingesting of drugs". The WHO definition expands on this by including outcome variables. Drug utilization is defined by WHO as "the marketing, distribution, prescription, dispensing, and use of drugs in a society, with special emphasis on the resulting medical and social consequences".

Drug utilization studies are a vitally important approach to establishing what medicines are used and how they are used. They can be implemented at the facility, regional, sub-sector, sector or international levels. In this section we consider the objectives, methods and uses of these studies.

12.2.1   Objectives of drug utilization studies

Drug utilization studies have four principal objectives. These are:

- to measure the use of pharmaceuticals;
- to understand the reasons for their use;
- to educate health professionals; and
- to design, monitor, and evaluate interventions aimed at improving the use of medicines.

12.2.2   Methods of drug utilization studies

There are both quantitative and qualitative methods available for drug utilization studies. Quantitative methods measure the situation under investigation, whilst qualitative methods help us to understand the reasons for medicine use problems.

Quantitative methods for drug utilization studies include:

- the recording of WHO indicators on medicines use,
- the collection of aggregate data from health care systems, organizations or health facilities on medicines consumption,
- case record reviews and prescription audits in health facilities, and
- surveys of health care providers or communities on specific aspects of the medicines use process.

Combinations of methods can be used to provide both quantitative and qualitative data. These include:

- focus group discussions with users or providers,
- in-depth interviews of key informants,
- structured observations of medicines use encounters (defined as the "contact period" between health care provider and user),

- questionnaires, and
- simulated patient surveys.

The last method is also known as the "mystery client method". It is a form of covert research whereby a researcher takes on the role of a fictitious patient, and reports the encounter. This approach is particularly useful to identify gaps between knowledge, attitude, and practice in health professionals. However, there are some obvious ethical dilemmas with this method, and it should only be used where it is clearly indicated.

12.2.3   Uses of drug utilization studies

The proposed medicine use indicators can be part of a routine health management information system, collected retrospectively in a survey of medical records, or collected prospectively in an observational study. Box 12.2 illustrates the WHO medicine use indicators.

The WHO indicators are divided into *core medicine use indicators* and *complementary medicine use indicators*. Core medicine use indicators are further divided into prescribing, patient care and health facility indicators. The health facility and complementary medicine use indicators in particular require additional information about the local health care system, which may influence the results.

These indicators can help to identify specific problems of medicines use, but they also allow the follow-up of interventions or to compare the medicines use situation in different facilities or health care systems. In the case of interventions it is particularly important to have baseline data before the intervention, to have a control group without the intervention, and to follow up changes over time, and especially after the intervention. Whilst the former two approaches are important to separate spontaneous and seasonal changes from improvements due to the intervention, the latter approach differentiates between short-term and long-term improvements. When facilities or systems are compared, it is important to take into account differences in staffing, equipment, epidemiology, socio-economic and cultural situation and other possible factors.

12.2.4   Case study-medicines utilization in Germany

A case study from Germany uses the ATC/DDD system to compare and analyse the aggregate prescription data of the statutory sickness funds. As these sickness funds cover about 90 per cent of the population, the data provide an overview of the medicine use situation and the changes from 1998 to 1999 in the German health care system. The saving potentials are analysed in terms of three elements: prescribing the least expensive generic medicines instead of more expensive generic or brand products; using less expensive but therapeutically equivalent medicines for certain indications such as essential hypertension or duodenal ulcer; and the substitution of

---

Box 12.2: WHO medicines use indicators

Core medicine use indicators

Prescribing indicators
1. Average number of medicine per encounter.
2. Percentage of medicine prescribed by generic name.
3. Percentage of encounters with an antibiotic prescribed.
4. Percentage of encounters with an injection prescribed.
5. Percentage of medicine prescribed from essential medicine list or formulary.

Patient care indicators
6. Average consultation time.
7. Average dispensing time.
8. Percentage of medicine actually dispensed.
9. Percentage of medicine adequately labelled.
10. Patients' knowledge of correct dosage.

Health facility indicators
11. Availability of a copy of essential medicine list or formulary.
12. Availability of key medicine.

Complementary medicine use indicators
13. Percentage of patients treated without medicine.
14. Average medicine cost per encounter.
15. Percentage of medicine costs spent on antibiotics.
16. Percentage of medicine costs spent on injections.
17. Prescription in accordance with treatment guidelines.
18. Percentage of patients satisfied with the care they received.
19. Percentage of health facilities with access to impartial medicine information.

Source: Adapted from WHO/DAP (1993) *How to Investigate Medicines Use in Health Facilities: Selected Medicines Use Indicators*. WHO/DAP/93.1. Geneva: World Health Organization.

---

medicines with unknown efficacy by medicines with proven effectiveness. These are illustrated in Box 12.3.

12.3   Pharmacoeconomics

Pharmacoeconomics is a relatively young discipline concerned with the study of how people and society end up choosing to employ scarce productive resources that could have alternative uses, to produce medicines and other pharmaceutical products, and distribute them amongst various people and groups in society to enhance quantity and quality of life. It analyses the costs and benefits of improving patterns of resource allocation in production, distribution and consumption of such products.

Box 12.3: **Saving potentials through prescription of generic medicines, cheaper therapeutically equivalent medicines and substitution of medicines of doubtful effectiveness in the German health care system**

Medicine group	Cost 1998 [Million DM]	Cost 1999 [Million DM]	Change [Million DM]
Generic medicines			
Total prescriptions	11.765,5	14.465,2	+ 2.699,7
Least expensive generic option	9.274,0	11.497,6	
Saving potential	2.491,5	2.967,6	+ 476,1
Therapeutically equivalent medicines			
Total prescriptions (limited to 17 therapeutic categories)	3.513,4	4.429,7	+ 916,3
Generic substitution (saving potential already included under generic medicines)	2.983,8	3.870,7	
Therapeutically equivalent chemical substitution	1.190,3	1.608,6	
Saving potential	1.793,2	2.262,0	+ 468,8
Medicines with unknown efficacy			
Total prescriptions	5.441,2	4.706,6	− 734,6
Substitution with effective medicines	1.992,0	1.712,1	− 279,9
Saving potential	3.349,2	2.994,5	− 354,7
Total saving potential	7.633,9	8.224,1	+ 590,2

Source: Adapted from Schwabe, U. and Paffrath, D. (2001) *Arzneiverordnungsreport 2000*. Berlin: Springer Verlag.

Pharmacoeconomics is both multi-disciplinary and multi-functional, and its scope is therefore very wide ranging. The narrowest definition of the subject would be to equate it with the economic evaluation of medicines. It is useful therefore to have a definition that locates its position within the management of pharmaceuticals in international health.

12.3.1   Definition of pharmacoeconomics

Pharmacoeconomics can be defined briefly as *"the discipline that describes and analyses the costs and benefits of pharmaceutical and alternative therapies to the health care system, the different stakeholders, and society as a whole"*. This definition indicates some of the dilemmas and difficulties of the topic. The costs are not simply those things that involve a monetary transaction, but include the use of all resources that are usually measured in monetary terms.

Costs and benefits can be both medical or non-medical, such as for transportation. They can comprise employment and productivity, and intangible benefits and costs, such as well-being and suffering. There are not only different types of costs involved but also different perspectives. Transport costs, for example, are a real cost from the patient's viewpoint, but may not be seen as a cost from the provider's viewpoint. As benefits and costs also occur at different points in time economists have developed a technique to calculate them at present prices with a discount rate for the future.

There is an important difference between a *private* and a *social* discount rate. In the first, a person or company usually chooses the market rate of interest that somebody has to pay to borrow money. A social rate of discount, on the other hand, is chosen by a government and should reflect the preference of society for present over future benefits. If the selected rate is high, quick financial returns become more attractive, and a long-term perspective is far less appealing. One practical consequence of this is the man-made extinction of plant and animal species. It is often argued therefore that governments should opt for a low social rate of discount.

Other factors need to be considered in a pharmacoeconomic evaluation. These are:

· the time horizon,
· the alternative (or comparator) therapy,
· the need for a sensitivity analysis for the different assumptions made in the study, and
· the need for an incremental analysis to assess the relative economic attractiveness in comparison to alternative therapies.

All these different considerations underline the tension that exists in pharmacoeconomics between methodologies (which try to take into account all factors) and transparent and robust approaches (which are simplified models of reality). The latter are more easily understood by potential users and less prone to manipulation by vested interests.

12.3.2 Methods of pharmacoeconomics

The basic pharmacoeconomic methodologies include four different approaches:

· cost minimisation analysis,
· cost effectiveness analysis,
· cost utility analysis, and
· cost benefit analysis.

Cost minimisation analysis is the most straightforward of these approaches, requiring the least information. The outcomes or benefits are proven to be equivalent, so that only the costs of different therapies need be examined and compared. However the costs are not confined to the price of medicines: they should also include the costs of preparation and delivery of treatments, of monitoring their use, and the costs of treating any adverse reactions.

Increasingly, societies have become interested in costs and cost-effectiveness of pharmaceuticals as another important criterion in the selection process in addition to safety, effectiveness, and quality. This additional criterion could either be applied at the stage of marketing approval, pricing decision, or reimbursement status of a pharmaceutical.

The introduction of a needs clause in Norway in the 1930s was an early example of the application of a national economic perspective to the licensing of a new medicine: however, this approach was abandoned in the 1990s because of the need to harmonize medicines regulation across Europe. But in Australia and Canada pharmacoeconomic considerations continue to play an important role in the pricing and reimbursement decisions of medicines.

A cost effectiveness analysis measures the costs in monetary terms, whilst the outputs are recorded in health improvements such as clinical cures or prevention of death. The obvious advantage of this approach is that benefits are measured in the units of relevant health outcome. This allows a comparison between therapies of similar outcomes. However, a therapy judged to be highly cost effective by society in terms such as a reduced inpatient stay may not be interpreted as a positive outcome by self-financing hospitals.

A cost utility analysis attempts to overcome the problem of multiple and different outcomes from certain therapies. Different outcomes are combined according to their desirability. Often this is expressed in the gain of quality-adjusted life years (QALYs). The basic unit of the QALY is the life year. Two treatments may extend life, but if these extra years are filled with pain in one case but pain free in the other the value of the treatment is different. Applying a fraction to the number of years increase in survival, which values these years relative to full health (1.0), makes a quality adjustment, and produces a figure for the treatment's impact in QALYs.

Another measure of health gain is the disability-adjusted life year (DALY). This measures the burden of disease in terms of premature death and disability, and is calculated in a similar way to the QALY.

In *cost benefit analysis* all costs and benefits of a therapy are expressed in monetary terms, and discounted at a specified interest rate to the present value. The net savings calculation of negative costs and positive benefits is compared between different therapies. The one with the highest net savings gives the best economic value.

Theoretically, this method allows the comparison of dissimilar outcomes, but these are of course inextricably linked to the transformation of different costs and benefits into monetary measurements.

12.3.3   Uses of pharmacoeconomics

Pharmacoeconomics plays an increasingly important role in health care systems as a tool for the analysis of the social costs of pharmaceutical care, and to identify saving potentials. The absolute costs of health care and its share of the gross domestic product have been rising in most high-income countries, from about 6 per cent in the 1960s to around 9 per cent in the 1990s. Pharmaceutical expenditure, in both relative and absolute terms, varies considerably between national health care systems. Examples of this variation are illustrated in Box 12.4.

In high income countries pharmaceuticals represent between 10 and 20 per cent of total health care expenditure, whilst in low income countries this may rise to 30 or 60 per cent. However, the absolute value of expenditure on medicines in low and middle income countries rarely exceeded US $ 20 per head in 1990. According to a more recent OECD report, expenditure on medicines has been rising rapidly in many high income countries since then. So the cost of pharmaceuticals is an important factor in the health care expenditure of all countries.

12.4 Pharmacovigilance

Pharmacovigilance is concerned with the safety of medicines following ingestion. Its focus is on adverse reactions to medicines. As we saw in chapter 10, an ADR is any response to a drug that is noxious and unintended and occurs at doses normally used for prophylaxis, diagnosis, or therapy. Pharmacovigilance is a structured process for the monitoring and detection of ADRs in a given context.

12.4.1 Definition of pharmacovigilance

Pharmacovigilance can be defined as "the science of postmarketing surveillance, evaluation, and signalling of the undesirable effects of pharmaceuticals, where the major sources of new information are spontaneous reporting of such effects".

Data derived from sources such as as Medicines Information, Toxicology and Pharmacovigilance Centres have great relevance and educational value in the management of the safety of medicines. Medicine-related problems, once detected, need to be assessed, analysed, followed up and communicated both to regulatory authorities, health professionals and the public. Pharmacovigilance includes the dissemination of such information. In some cases medicines may need to be recalled and withdrawn from a market, a process that entails concerted action by all those involved at any point in the medicines supply chain.

Pharmacovigilance therefore includes all the regulatory measures that are required in order to prevent future ADRs and to improve the benefit/risk ratio of pharmaceuticals: for example, package inserts may have to be adapted. Pharmacovigilance thus enables effective decision-making by national medicines regulatory authorities, as well as facilitating international collaboration, with the objective of ensuring effective and safe use of medicines.

Box 12.4: Pharmaceutical consumption per capita in selected countries (1975/1990/2000*)

Country	1975	1990	2000
High income			
Australia	60.1	87.6	291.4
Canada	55.2	124.0	405.1
France	108.8	223.3	486.9
Germany	114.8	221.4	378.1
Norway	52.2	89.2	201.8 (1997)
Switzerland	115.2	166.4	338.1
United Kingdom	53.5	97.4	239.5 (1997)
United States of America	90.4	190.6	540.3
Low and middle income			
Afghanistan	1.8	1.2	n/a
Iran	24.8	37.2	n/a
Brazil	18.6	15.6	n/a
Mexico	25.3	27.7	96.2
Bangladesh	1.5	1.6	n/a
China	6.4	7.1	n/a
India	1.6	3.3	n/a
Philippines	9.8	11.4	n/a
Botswana	6.7	20.1	n/a
Ghana	10.1	10.0	n/a
Tanzania	5.5	4.0	n/a

Notes: Whilst the figures for 1975 and 1990 are presented in constant 1990 US $, those for 1997 and 2000 are actual figures from OECD statistics. (*OECD Health Data* 2003. Paris: Organization for Economic Cooperation and Development).
n/a = not available
Source: Adapted from *Global comparative pharmaceutical expenditures with related reference information* (2002) EDM/PAR/2000.2. Geneva: WHO.

12.4.2 Methods of pharmacovigilance

Lay people often assume that most serious ADRs will have been detected before a new pharmaceutical is introduced to the market. However, this is highly unlikely given the design of pre-marketing studies. There are three phases to such studies:

· *Phase 1 studies* evaluate the safety of a pharmaceutical in a small number of healthy human beings.
· *Phase 2 studies* aim to define an effective dose in a small group of closely monitored and carefully selected patients.

· *Phase 3 studies* are conducted in a few thousand carefully selected patients to assess safety and efficacy of the pharmaceutical.

Dose-dependent and pharmacologically predictable ADRs can therefore be identified with these studies, which are also called type A ADRs (see chapter 10). However, adverse reactions with a frequency of less than 1 in 1,000, or which are due to long-term therapy, will not be detected. Moreover, it is unlikely that the more serious, unpredictable and rare type B ADRs (which are due to hypersensitivity or an idiosyncratic mechanism, chapter 10) will be detected.

In addition, there are other limitations due to the discrepancy between the well-controlled conditions of phase 1 to 3 studies and the use of a pharmaceutical in the patient population of everyday clinical practice. In these circumstances patients often handle pharmaceuticals differently, have particular demographic characteristics such as pregnancy, low or high age, have concurrent diseases, or take concurrent pharmaceuticals. All of these may influence the development of an ADR in a patient.

There may be significant differences in the occurrence of medicines-related problems amongst countries or regions due to differences in local circumstances. Some of these are listed in Box 12.5.

12.4.3 Post-marketing surveillance

Pre-marketing trials do not only lack the power to detect important ADRs which may occur at rates of 1 in 10,000 or fewer medicines exposures, but also lack the capacity to detect ADRs widely separated in time from the original use of the medicines, or delayed consequences associated with long-term medicines administration. These trials often do not include special populations such as pregnant women, the elderly or children who may be at risk for unique ADRs or for an increased frequency of ADRs compared with the general population.

Post-marketing surveillance, sometimes called phase 4 clinical studies, is an important tool to detect less common but sometimes serious ADRs. The data and information from preclinical and clinical testing, the pre-marketing phase, is inevitably incomplete. Principal factors contributing to this are:

· animal testing cannot sufficiently predict human safety,
· the patient population in clinical trials is selected and limited in number,
· the disease conditions differ from those in real clinical practice, and
· the duration of trials is limited.

The spontaneous reporting of ADRs is the cornerstone of post marketing safety monitoring of pharmaceuticals. This reporting can be done to the regulatory authority of a country, or to the pharmaceutical company concerned. In most countries companies are obliged to pass the information on to the regulatory authority. Companies may also have to submit international reports, giving details of adverse reactions reported in other countries, to the national authority.

> **Box 12.5: Factors contributing to differing levels for ADRs between countries**
>
> · Diseases and treatment practices.
> · Genetics, diet, traditions and health behaviour.
> · Medicines manufacturing processes and quality of medicines.
> · Medicines use and treatment guidelines including indications and doses.
> · Medicines distribution and supply affecting availability of medicines.
> · Regulatory policies and practices affecting availability and banning of medicines.
> · Use of traditional and complementary medicines such as herbal medicines.
> · Presence of counterfeit and substandard medicines.

In certain countries only doctors are allowed to report suspected ADRs, whilst in others reports from a wider group of health professionals are accepted. These may include pharmacists and nurses. The frequency of reporting varies considerably between countries. It is estimated that only between 1 and 10 per cent of all cases are actually reported to the authorities.

12.4.4   Uses of pharmacovigilance

One reason for this serious under-reporting is the difficulty that health professionals have in recognizing that an adverse event in a patient is linked to a certain medicine, and is therefore an ADR. This is particularly the case, if

- an adverse event occurs frequently in the untreated population as so-called "background noise",
- there is a long latent period between medicines-taking and the adverse event, or
- the adverse event is not usually linked to medicines-taking.

Other problems of spontaneous reporting are the lack of control groups with comparative information and the unknown denominator for a reported ADR (how many treatment episodes have been completed). Epidemiological methods are therefore increasingly used for post-marketing pharmaceutical studies.

12.5   Pharmacoepidemiology

The methods of pharmacovigilance are based on epidemiology. The United Kingdom set up the first national pharmacovigilance system, known as the Committee on the Safety of Medicines (CSM) in 1968. One of the pioneers of this so-called "yellow card system", Professor Bill Inman, later developed the prescription event monitoring system (PEM). This was a cohort approach designed to monitor adverse events of newly licensed medicines in order to identify possible ADRs.

In the United Kingdom, the method involves the following up of the first ten thousand prescriptions of a new medicine. Patients serve as their own controls when they stop taking the pharmaceutical product. However, one of the major limitations of this method is the limited response rate of medical practitioners: it is only around 50 per cent.

12.5.1   Definition of pharmacoepidemiology

Pharmacoepidemiology can be defined as "the application of epidemiological knowledge, methods, and reasoning to the study of the effects (both beneficial and adverse) and uses of drugs in human populations". Its objectives are to describe, explain, control, and predict the effects and uses of pharmacological treatments in a defined time, space and population.

Epidemiology studies the distribution of health and disease in human populations. Since medicines are amongst the factors that influence such distribution pharmaco-epidemiological studies are important. Pharmacovigilance is one application of pharmacoepidemiology: the application of epidemiological methodologies to the effects and uses of medicines is most clearly demonstrated in post-marketing surveillance.

12.5.2   Methods of pharmacoepidemiology

Other epidemiological methods applied to the safety of medicines are case reports and case-control studies. These offer the advantage that rare ADRs can be detected at low cost and relatively high speed in comparison to cohort studies. They are prone to biases, cannot determine causation unless the adverse event is very rare, and do not permit measurement of the incidence of an ADR.

12.5.3   Uses of pharmacoepidemiology

Whilst the origins of pharmacoepidemiology were firmly linked to pharmacovigilance and the study of the safety of medicines, today the discipline includes all issues of pharmaceuticals and their interaction with people. The uses of pharmacoepidemiology can be illustrated by listing the kinds of questions that pharmacoepidemiology can be asked to address. Examples of these are listed in Box 12.6.

Other approaches to the investigation of the uses of medicines, including pharmacoeconomics, medicines utilization reviews, and health systems and health policy research concerned with pharmaceuticals are also closely allied to pharmacoepidemiology. Indeed, a more appropriate term summarising these approaches might be "public health pharmacology", as discussed in chapter 3, as the overall research field is the study of pharmaceuticals for health and disease in human populations.

Box 12.6:   Examples of questions that can be addressed by pharmaco-epidemiology

---

· Are there differences in the number of patients diagnosed and treated with a particular condition amongst different populations within a country?
· How much of the past year's decline in incidence of a particular disease can be accounted for by the effects of specific medicines?
· What is the effectiveness of a particular medicine in a defined population?
· To what extent does that effectiveness depend on age, gender, and socio-cultural level?
· What changes can we predict in the prevalence of a disease based on current medicines consumption trends?
· What are the most common uses of a particular medicine?
· To what extent do these uses follow current clinical recommendations and guidelines?
· What factors should guide the decision to conduct formal post-marketing epidemiological studies?
· How can we accelerate the discovery of new clinically relevant adverse medicines reactions?
· How should the validation of large clinical databases be approached?

Source: Porta, M. and Hartzema, A. (1987) "The contribution of epidemiology to the study of medicines". *Medicines Intelligence and Clinical Pharmacy* 21: 741–747.

12.6   Health systems and policy studies

An important distinction needs to be made between *health systems* and *health care systems* (sometimes called health services). Health systems comprise all the interactions, institutions and organizations that are linked to health in any way. Health care systems consist of all the structures, processes, outputs and outcomes of health care services.

· The goal of *health systems research* is the health of the public, and its aim is to improve the effectiveness and efficiency of the health system of a society.
· *Health policy studies* add the dimension of actions or intended actions to modify the institutions and organizations of the health system in a society.
· *Health services research* is concerned with all aspects of the structure, process, output and outcomes of health care services.

12.6.1   Levels of analysis

Important characteristics of these studies are their operational link between research and development and their interdisciplinary design. They can be conducted at one of three levels in the organization:

· the *micro level* of a unit within the health care system;
· the *meso level* such as a hospital or district; or
· the *macro level* of society as a whole.

12.6.2   Regulatory science

A specific category of health systems research linked to pharmaceuticals is *regulatory science*. This deals with the control of pharmaceuticals in society and the role of the different stakeholders such as the pharmaceutical industry, the regulators and various citizens' organizations.

Some important regulatory issues include:

· the public health perspective;
· the accessibility of information;
· the transparency and democratic accountability of decision-making;
· the independence of regulators and scientists from commercial interests; and
· the systems of pharmaceutical safety testing and pharmacovigilance in a country.

An important initiative in regulatory science relates to the supra-national harmonization of medicines regulation in the European Union. This has been analysed in detail by Abraham and Lewis, 2000 (see further reading). It provides an interesting case study as to whether the new harmonised approach favours public or commercial interests.

12.7   Conclusion

In this chapter we have considered the main research methods available to us in obtaining new information about the use of medicines. Obtaining this information is absolutely vital. Without it we have little way of knowing whether medicines are reaching the people who need them, whether they are being used appropriately where they are available, and whether we are making progress in improving the health of populations by means of pharmaceuticals.

This brings the process of managing pharmaceuticals in international health to an end. We have looked at ways of improving access to medicines, and we have looked at methods for ensuring their rational use once they are available. We have looked at some of the problems associated with medicines use, such as adverse medicines reactions and the development of resistance to antibiotics, and we have considered information that is available to us about the actions and uses of those medicines. Finally, in this chapter, we examined some of the tools available to us to see what effect this is having.

In our next chapter we review trends and prospects for the future in relation to the goal of achieving greater access to pharmaceuticals by the world's poor. There are some trends that appear to represent a real threat to greater access and availability of medicines: on the other hand a number of recent developments seem to offer hope and encouragement that it may just be possible to make a significant impact on access, and to improve health and well-being worldwide, in the future.

Further reading

Abraham, J. and Lewis, G. (2000) *Regulating Medicines in Europe: Competition, Expertise and Public Health.* London: Routledge.

Bowling, A. (2002) *Research Methods in Health: Investigating Health and Health Services.* Buckingham: Open University Press.

Cobert, B.L. and Biron, P. (2002) *Pharmacovigilance from A to Z: Adverse Medicines Event Surveillance.* Oxford: Blackwell Science.

Dickson, M. and Bootman, J.L. (1996) "Pharmacoeconomics: an international perspective". In: *Principles of Pharmacoeconomics*, Bootman, J.L., Townsend, R.J. and McGhan, W.F. (eds.) pp 20–43. Cincinnati: Harvey Whitney Books Company.

Dukes, M.N.G. (1993) *Medicines Utilization Studies: Methods and Uses.* Geneva: World Health Organization.

Inman, W. (2003) *Don't Tell the Patient: Behind the Drug Safety Net.* Bishop's Waltham: Highland Park Productions.

International Drugs Surveillance Department (1991) *Medicines Safety: A Shared Responsibility.* Edinburgh: Churchill Livingstone.

Management Sciences for Health: International Network for Rational Use of Drugs (1995) *How to Use Field Methods to Design Drug Use Interventions.* Boston: MSH/INRUD.

Quick, J.D., Rankin, J.R., Laing, R.O., O'Connor, R.W., Hogerzeil, H.V., Dukes, M.N.G. and Garnett, A. (eds.) (1997) *Managing Drug Supply: The Selection, Procurement, Distribution and Use of Pharmaceuticals.* Second edition. West Hartford, CT, USA: Kumarian Press.

Schulman, K.A., Glick, H.A., Polsky, D., John, K.R. and Eisenberg, J.M. (2001) "Pharmacoeconomics". In: *Medicines Benefits and Risks, International Textbook of Clinical Pharmacology*, van Boxtel, C.J., Santoso, B. and Edwards, I.R. (eds.) pp 37–53. Chichester: Wiley & Sons Ltd.

Strom, B.L. (2000) (ed.) *Pharmacoepidemiology.* Third edition. Chichester: Wiley.

WHO/DAP (1992) *How to Investigate Drug Use in Communities.* WHO/DAP/92.3. Geneva: World Health Organization.

WHO/DAP (1993) *How to Investigate Medicines Use in Health Facilities: Selected Medicines Use Indicators.* WHO/DAP/93.1. Geneva: World Health Organization.

Chapter 13
Trends and Developments

Stuart Anderson and Karin Wiedenmayer

Box 13.1: Learning objectives for chapter 13

By the end of this chapter you should be able to:

· Describe the contribution of ethno-pharmacology to the development of new medicines.
· Define and distinguish between pharmacogenetics and pharmacogenomics.
· Describe the implications of pharmacogenomics for international health.
· Define e-commerce and list its advantages over traditional forms of commerce.
· List the disadvantages of e-commerce relating to the supply of medicines.
· Define pharmaceutical care and medicines management.
· List the steps involved in pharmaceutical care.

13.1   Introduction

In these last two chapters we aim to bring together a number of the strands in this book, to consider current themes in the management of pharmaceuticals, and to reflect on prospects for the future. This complex scene is changing rapidly: a host of factors are having an impact and are likely to play a significant part in shaping the future. In this chapter we focus on the impact of science, technology and workforce issues: in the next we focus on policy initiatives and their implications.

We begin by looking at the prospects for the continuing use of natural resources for medicinal purposes: we consider the contribution of herbal remedies, as well as some of the problems created by their increasing popularity in industrialised countries. We consider the likely impact of science and technology on the development of new medicines: in particular, we explore the prospects that are opened up by the new sciences of pharmacogenetics and pharmacogenomics, and what contributions these might make to international health.

We move on to the role of the internet, and both the opportunities and problems that are created by it as far as medicines are concerned. We then return to some of the issues concerning health professionals, and particularly the increasing focus on medicines management and pharmaceutical care. We return to the wider policy issues in the final chapter.

13.2   Trends in the use of herbal remedies

Despite the shift towards Western medicines, herbal remedies remain the foundation of much therapy in developing countries. At the same time their popularity continues to increase in many developed countries. The global trade in herbal remedies is now worth US $ 18 billion per year. It is estimated that up to 50,000 plants are used for medicinal purposes somewhere in the world. Although some of these are cultivated commercially, the vast majority are not. Most are simply taken from their natural habitats. The result is that up to one fifth of these plants, around ten thousand, are now under threat.

13.2.1   Ethnopharmacology

As we have seen in chapter 4, traditional healers and pharmacists in developing countries are an important source of information about plant sources for new medicines. Only a fraction of the earth's natural pharmacopoeia has been analyzed with modern techniques. The threat of imminent extinction of many plant species, especially in tropical areas, has encouraged scientists to investigate this source of potential pharmacological agents. This process of *ethnopharmacology* requires the observation and recording of medical techniques, identification of plant materials, and experimental investigation of the ingredients and their effects. Box 13.2 illustrates a recent example in which a herbal remedy has been shown to have important pharmacological actions.

Although developing countries provide many of the raw materials needed in medicine manufacturing, the final products are often returned as high-priced medicines. As more plants are needed for large-scale production, over-harvesting has led to stock depletion. As we have noted, traditional medicines play an important role in the provision of health care in developing and developed countries, increasing their commercial value.

13.2.2   The patenting of traditional remedies

There have been a number of cases of patenting of traditional medicines without consent from, or compensation to, those who introduced them. This has focused further attention on their importance. Traditional medicine usually involves biological resources and the knowledge of local and indigenous peoples and/or healers regarding their medicinal use. Traditional medicine is therefore closely connected with biodiversity, conservation and indigenous peoples' rights over their knowledge and resources. This situation raises complex ethical and legal questions and dilem-

---

Box 13.2: Chinese herbal tea and jaundice

West sees cure for jaundice in tea leaves

A Chinese herbal tea used in Asia to treat jaundice in newborn babies really works, an American team has found. Better still, they know why.

Many herbal remedies contain powerful ingredients and there is no doubt that some, at least, are effective. Western medicine has not adopted them because they contain such a mass of different components that it is difficult to know which is the most important. They cannot be patented, either, so there is little economic incentive to carry out trials.

A team at the Baylor College of Medicine in Houston, Texas, studied a herbal tea called Yin Zhi Huang, made from *Artemisia capillaris* and three other herbs. Teas with these ingredients are prescribed for jaundice by practitioners of traditional Chinese medicine.

Experiments showed that the tea's active ingredient is a compound called 6,7-dimethyl-esculetin; when synthesised chemically, it has the same effect-making it the vital substance in the tea.

The team at Baylor, led by Dr David Moore, have identified a protein on the surface of liver cells as a key regulator of bilirubin clearance. The more active that this constitutive androstane receptor (CAR) is, the more quickly bilirubin in cleared.

The team report in the *Journal of Clinical Investigation* that in mice, three days of feeding with herbal tea accelerated the clearance of bilirubin injected into the bloodstream. The same applied to mice genetically engineered to carry the human CAR gene, but not to those lacking the gene. That showed that CAR is important in clearing bilirubin, and that herbal tea improves its performance.

In the journal Dr Mitchell Lazar, of the University of Pennsylvania School of Medicine, calls the research 'a wonderful example of knowledge gained by applying the Western scientific method to an Eastern herbal remedy'.

Source: The Times, London, 3 January 2004.

---

mas concerning intellectual property ownership and the protection of traditional medicine. It remains a major challenge for the future.

13.3   The contribution of pharmacogenetics and pharmacogenomics

One of the greatest areas of impact on healthcare in the future, in both developed and developing countries, is likely to come from the fields of pharmacogenetics and pharmacogenomics. The classical model of pharmacotherapeutics has been that 'one medicine fits all patients'. Developments in pharmacogenetics mean that this approach will increasingly give way to more individualised therapy, in which the patient's unique genetic make-up is matched with an optimally effective medicine.

The variation in genes between different members of a population means that they produce different forms of proteins in different quantities. These proteins include some that metabolise medicines or are a site of drug action. The result of this variation is that different patients can respond differently to the same medicine. Some may not respond to a specific medicine, whilst others may experience serious adverse reactions. Measuring the DNA differences between individuals can thus help to predict the variation in response to any given medicine.

13.3.1 Pharmacogenetics

Pharmacogenetics attempts to examine the underlying reasons for the differences in drug response. These can include:

- *Genetic differences in medicine metabolism*: Some patients metabolise certain medicines much more quickly than others. This variation is due to genetic differences in how their liver enzymes work. Such patients will need to take higher doses of medicines to obtain the necessary blood concentration, whilst those who metabolise the medicine more slowly will require lower doses in order to avoid accumulation of the medicine and toxic effects.
- *Genetic predictors of response*: Some patients fail to respond to the medicine at all. One reason for this might be that they have a variant of the disease that is triggered by a different genetic mechanism to the one for which the medicine was designed.

13.3.2 Pharmacogenomics

One of the great scientific achievements of recent years has been the decoding of the human genome. This work is now being extended to the genomes of infectious organisms. The deciphering of these genomes will not of itself provide cures for all known diseases, but it will provide very different ways of identifying, preventing and treating disease in the future to those we use today.

Pharmacogenomics is the application of this knowledge of the human genetic code to the design of medicines. It aims to establish a signature of DNA sequence variants that are characteristic of individual patients, in order to assess disease susceptibility and select the optimal medicine treatment. It involves:

- the study of "genetic polymorphisms" in families and populations;
- the techniques of population genomics and epidemiology; and
- the search for disease-related genes.

The distinction between pharmacogenetics and pharmacogenomics is somewhat blurred. However, pharmacogenomics is concerned not only with the genetic variability in drug response but also with the search for the genes responsible. It also plays an important part in identifying new gene targets for which specific drug treatments can be tailored. Before the genome project scientists had about four hundred specific targets at which to direct medicines: so far, the human genome project has presented them with another ten thousand. How significant this is remains to be seen.

13.3.3 Implications for international health

The hope is that ongoing research in pharmacogenetics will help make medicines more effective. What it will not do is make them more accessible and more affordable for those in low and middle income countries. Indeed, this type of research will be very expensive, very complex and take many years. Those who invest in it will expect a return on their investment. Under the "trickle down" hypothesis we might reasonably expect the cost of this technology to fall, such that eventually the benefits reach those for whom access to even the most basic of medicines remains an aspiration. However, experience over the last twenty years suggests that this is not likely without structural changes in the world economic system.

Nevertheless, in industrialised countries the pace of development is fast. In the not too distant future it is likely that patients will be able to undergo a genetic test to predict their response to particular medicines. This approach should contribute substantially to improved clinical outcomes, reduce wastage and help avoid serious side effects. New pharmaceutical products linked to pharmacogenetic tests are likely to become available within the next five years. New technology is currently being tested that should allow pharmacogenetic tests to be carried out quickly in primary care settings.

Whether even developed countries will be able to afford these developments is another matter. Recent data from the OECD demonstrate that health spending in many industrialised countries is rising faster than economic growth. This is largely due to growing expenditure on health care technology and new medicines. Just how long developed countries will be able to sustain technological developments at their current rate remains to be seen. Examples of how this growth has increased are illustrated in Box 13.3.

So pharmacogenetics and pharmacogenomics offer more hope for the future than solutions to the more immediate and pressing problems described in this book. They offer the prospect that at least some of the "orphan diseases" will no longer be neglected, although equally for others the costs of finding solutions may make them even less likely to be the subject of extensive research. Indeed, even with the major diseases there are likely to be problems. Tiny variations in the genome mean that there will always be a few individuals who cannot respond to the treatments that help the many.

13.4   The changing role of the internet

The establishment of the internet and the world wide web have opened up enormous opportunities for extending commerce and the provision of information. They have already had a significant impact on low and middle income countries, and

**Box 13.3:   Percentage of health expenditure spent on pharmaceuticals –
selected OECD countries 1980 to 2000**

Country	Year		
	1980	1990	2000
Australia	8.0	9.0	12.4
Canada	8.5	11.5	15.7
Denmark	6.0	7.5	8.7
Finland	10.7	9.4	15.5
Germany	13.4	14.3	13.6
Greece	18.8	14.3	14.2
Ireland	10.9	12.2	10.6
Japan	21.2	21.4	15.9
Luxembourg	14.5	14.9	12.1
Netherlands	8.0	9.6	10.1
Sweden	6.5	8.0	13.9
United States	9.1	9.2	11.9

Note: Figures give total expenditure on pharmaceuticals and other medical non-durables, as percentage of total expenditure on health for twelve selected OECD countries for the years 1980, 1990 and 2000.

Source: Adapted from *OECD Health Data 2003* (2004) Second edition. Paris: Organization for Economic Cooperation and Development.

offer both opportunities and threats in relation to the management of pharmaceuticals. Internet sales of pharmaceuticals are increasing and pose a new challenge for medicine regulation and quality assurance.

13.4.1   E-commerce and Electronic Data Interchange

E-commerce has been defined as "the sharing of business information, maintaining business relationships, and conducting business transactions by means of telecommunications networks". It began in the 1960s with satellite and cable links, moving on to Electronic Data Interchange (EDI) between computers in the 1970s, and the use of the world wide web in the 1990s. Global e-commerce now embraces almost every country, although the extent of its usage varies widely amongst the world's populations.

EDI has been defined as "the electronic communication of messages such as orders and invoices between companies' computers in such a way that the messages can immediately be actioned by the receiving company's systems without the need for manual keying". Technically, e-commerce is a more sophisticated form of EDI. It is based on a worldwide network of digital communications, which involve activities in many different branches of commerce and amongst different economic agents, such as pharmaceutical suppliers and health providers.

13.4.2   Advantages of e-commerce

The advantages of e-commerce are numerous. The main ones are:

- *Accessibility*: On-line shops are accessible around the clock.
- *Transparency*: The possibility to compare prices enhances transparency and competition.
- *Low prices*: Prices are generally lower than those of traditional shops due to reduced overhead costs.
- *Information*: Products are often showcased in graphic detail, and related information is offered.
- *Convenience*: E-shopping is convenient, time-saving and can be performed in the privacy of the customer's own home.

E-commerce in relation to pharmaceuticals is now well established in many parts of the world, and is now an accepted part of overall procurement by health service providers. The savings to be made from procurement in this way are substantial: in Great Britain e-procurement is estimated to generate savings of between 20 and 30 per cent in the US $ 10 billion NHS hospital procurement market. On-line pharmacies offer everyone the opportunity to purchase medicines. However, the usual barriers to access remain; internet prices will be unaffordable for most people, and even when ordered geographical and physical obstacles, such as lack of roads, still have to be overcome.

13.4.3   Disadvantages of e-commerce

However, e-commerce of pharmaceuticals poses some major problems. Firstly, as with other products advertised on the internet, the sales aspect generally comes first: this characteristic compromises the completeness and accuracy of the information offered, particularly with regard to prescription only medicines. Secondly, the internet provides a mechanism for the trafficking of illegal drugs: and thirdly, there is a lack of privacy and confidentiality on the internet. We will consider each of these matters in turn.

Firstly, professional control and supervision become crucial where prescription-only medicines are involved. Some on-line pharmacies operate in a geographically limited area and employ couriers with a basic medical education who check prescriptions on delivery. Larger sites establish contact with the prescribing doctor as well as the patient, which places an additional workload on doctors, and limits the free choice of a pharmacy. Other on-line pharmacies have no professional control in place, and even advertise the fact that this gap exists.

Some sell products like cannabis and cannabis seeds without any control, presenting them as harmless products. In 1999, in the first action of its kind, the US state of Kansas took legal action against companies selling prescription-only medicines on the internet without the buyers being examined by a doctor or obtaining a prescription. A year later, President Clinton proposed the drafting of legislation to regulate the growing sale of prescription medicines over the internet.

Secondly, the internet offers the prospect of increased drug trafficking. In 2001, a survey by the International Narcotics Control Board pointed out that internet drug trafficking had only recently come to the notice of most national authorities, and very few had taken legal action to stem it. National regulatory restrictions can be circumvented via the internet quite easily whilst there is no clear legislation in the field.

Thirdly, privacy and confidentiality are not guaranteed on the internet. Whilst special protection is often available for financial transactions, other electronic communication is easy to intercept. Possibly, internet users feel protected by the apparent anonymity of electronic communication.

13.4.4   Medicines information on the internet

Not only psychotropic and habit-forming medicines, but most other medicines can be potentially dangerous if they are sold without proper instruction and warnings. Sildenafil (Viagra), frequently offered online, can be potentially fatal if taken concurrently with nitrate products. The lack of communication with patients is another problem in the e-commerce of pharmaceuticals.

On-line pharmacies try to simulate advisory services by offering web-pages with articles on common problems, links to specialists or top selling products lists, or even decision-support systems. The more complex and informative these systems are, the more difficult they can be for patients to use, as they require an accurate and technical description of symptoms. In many cases, on-line advice cannot replace individual communication.

13.4.5   Strategies for safeguarding e-prescriptions

A number of strategies have been proposed to safeguard prescriptions that are sent electronically. For example, the International Pharmaceutical Federation (FIP) has issued guidelines aimed at securing safeguards in the use of electronic prescriptions. These are given in Box 13.4.

13.5   Changes in the healthcare workforce

Developments emerging from science and technology will have a significant impact on health professionals that is so far impossible to quantify. What is clear is that

> ## Box 13.4: FIP's Statement of professional standards on electronic prescriptions
>
> · A prescription must, as a minimum, contain information on patient identity, age and gender, medicinal product, strength, dosage and quantity, directions for the patient, and prescriber identity.
> · The system must allow the patient to ensure that the prescription is directed to the pharmacy of his or her choice.
> · The system must give the pharmacist access to such information about the patient as is necessary to enable the pharmacist to judge the correctness and appropriateness of a selected medicinal therapy.
> · Systems that collect or manage data from prescribing and dispensing activities, to be used for commercial purposes, must guarantee patient and prescriber confidentiality and that the dispensing pharmacy and pharmacist cannot be identified.
> · Systems must prevent third parties from interfering with the content of the prescription.
> · Systems must allow for authentication of prescriber and pharmacist identity.
> In addition,
> · Systems should provide for inclusion of diagnosis and/or intended use when desirable within a healthcare system.
> · Systems should, where appropriate, include provision for confirming an individual's entitlement to benefits within a health care scheme.
>
> *Source*: FIP website (2003). http://www.fip.org/resources/e-prescription-English.htm

pharmaceuticals will continue to play a vital part in improving the health of populations throughout the world for many years to come. The role of health professionals in managing this activity, in ensuring the availability of safe, effective medicines and their rational use, will be as important as ever.

But demographic factors are also having a major impact on the available health workforce. In developed countries this is largely a consequence of an aging population, with fewer young people coming forward to train as nurses and fulfil other key roles. In developing countries the AIDS epidemic is having a devastating effect on the availability of health workers.

For low and middle income countries, for whom the availability of sufficient numbers of properly trained health professionals is already a major problem, there continue to be major threats. Whilst real efforts are being made to increase the numbers of locally trained health professionals, the migration of such individuals from developing to developed countries seems set to get worse rather than better, unless far greater incentives are found for them to remain in their home countries.

13.5.1   Availability of pharmaceutical expertise

Some indication of the discrepancies in the numbers of health professionals available can be seen from figures for the number of pharmacists per 100,000 popula-

Box 13.5:   Number of registered pharmacists per 100,000 population for 25 selected higher income countries

Country	Number of registered pharmacists	Population (000)	Pharmacists per 100,000 population
Finland	7,500	5,172	145
Belgium	11,145	10,249	138
Iceland	340	279	122
Spain	29,820	39,910	119
France	62,800	59,238	106
Italy	60,340	57,530	105
Japan	132,180	127,096	104
Republic of Korea	45,340	46,740	97
Greece	8,920	10,610	84
Ireland	2,970	3,803	78
Portugal	7,165	10,016	76
United Kingdom	43,370	59,415	73
United States	201,095	283,230	71
Luxembourg	305	437	69
Sweden	6,100	8,842	69
New Zealand	2,460	3,778	65
Canada	19,070	30,757	62
Australia	11,485	19,138	60
Germany	51,050	88,017	58
Austria	4,440	8,080	55
Poland	21,235	38,605	55
Denmark	2,450	5,320	46
Hungary	4,385	9,968	44
Norway	1,925	4,469	43
Czeck Republic	4,315	10,272	42

Source: Anderson, S. (2002) "The state of the world's pharmacy: a portrait of the pharmacy profession". *Journal of Interprofessional Care* 16 (4): 391–404.

tion. Box 13.5 gives the number of pharmacists per 100,000 population for 25 developed countries. The number varies from 42 in the Czech Republic to 145 in Finland. Box 13.6 gives the number of pharmacists per 100,000 population for 25 low and middle income countries. These vary from around thirty in India to less than two in countries on the United Nations list of least developed countries.

These figures themselves disguise huge variations in the knowledge, training and practice of those involved. It is nonetheless clear that these countries will have a huge

Box 13.6:   Number of registered pharmacists per 100,000 population for 25 selected middle and lower income countries

Country	Number of registered pharmacists	Population (000)	Pharmacists per 100,000 population
Slovenia	695	1,988	35
Turkey	22,870	66,668	34
Jamaica	860	2,576	33
Belarus	3,230	10,187	32
Azerbaijan	2,505	8,041	31
India	300,000	1.008,937	30
Singapore	1,135	4,018	28
Thailand	15,478	62,806	25
South Africa	10,000	43,309	23
Chile	3,000	15,211	20
Bulgaria	1,315	7,949	17
Malaysia	3,560	22,218	16
TFYR Macedonia	320	2,034	16
Tajikistan	730	6,087	12
Bosnia and Herzegovinia	440	3,977	11
Georgia	438	5,262	8.3
Romania	1,600	22,438	7.1
Russia	9,340	145,491	6.4
Krgyzstan	275	4,921	5.6
Armenia	136	3,787	3.6
Uzbekistan	755	24,881	3.0
Albania	85	3,134	2.7
*UR of Tanzania	850	35,119	2.2
*Eritrea	53	3,659	1.4
*Gambia	10	1,303	0.8

*Included in United Nations list of 48 least developed countries
Source: Anderson, S. (2002) "The state of the world's pharmacy: a portrait of the pharmacy profession". *Journal of Interprofessional Care* 16 (4): 391–404

lack of capacity in pharmaceutical expertise for years to come, and that reliance on traditional practitioners will continue to play a major part in the health care of many populations. One of the challenges therefore is how to make greater use of such resources and how to integrate them better into local primary health care.

As we have seen in chapter 4, many countries have now opened schools of pharmacy, and indeed today there are over nine hundred schools of pharmacy worldwide. However these vary widely in the range and type of education provided. Since

1995 an Official World List of Pharmacy Schools (WLPS) has been published by the International Pharmaceutical Federation in collaboration with the International Pharmaceutical Students Federation, and can be accessed through the Cardiff University website (see further reading).

13.5.2   Pharmaceutical care

Two concepts have come to play an important part in both policy formation around the future role of pharmacists and in pharmaceutical education. These are pharmaceutical care and medicines management. We will now explore these concepts further.

Pharmaceutical care is delivered at the individual patient level. The definition most commonly quoted today is that attributed to Hepler and Strand: "pharmaceutical care is the responsible provision of medicine therapy for the purpose of achieving definite outcomes that improve a patient's quality of life". In 1998, the International Pharmaceutical Federation adopted this definition, with one significant change. In the revised definition "pharmaceutical care is the responsible provision of medicine therapy for the purpose of achieving definite outcomes that improve *or maintain* a patient's quality of life". This is seen as a more realistic goal, particularly for chronic progressive diseases where maintenance of quality of life would be a significant achievement.

The key words here are "responsible provision" and "definite outcomes". Whether a pharmacist is reviewing a prescription or a patient medication record, talking to a patient or responding to symptoms, he is automatically assessing needs, prioritizing and creating a plan to meet those needs. What has often been neglected is the need to adequately document, monitor and review the care given. These latter steps are essential to the practice of pharmaceutical care.

Pharmaceutical care implies the pharmacist's responsibility to the patient for the prevention of medicine-related illness. The pharmacist evaluates a patient's medicine-related needs, determines whether one or more medicine therapy problems already exist or may arise, and works with the patient and other health care professionals to design, implement and monitor a care plan. This plan aims to resolve medicine therapy problems and prevent potential problems. A "medicine therapy problem" can be defined as "an undesirable event, a patient experience that involves, or is suspected to involve medicine therapy, and that actually or potentially, interferes with a desired patient outcome".

Pharmaceutical care involves the following four steps:

Step 1: Assess the patient's medicine therapy needs and identify actual and potential medicine therapy problems
Step 2: Develop a care plan to resolve and/or prevent the medicine therapy problems
Step 3: Implement the care plan
Step 4: Evaluate and review the care plan

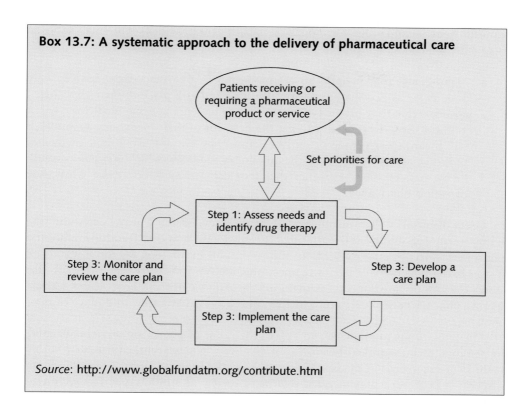

Box 13.7: A systematic approach to the delivery of pharmaceutical care

Patients receiving or requiring a pharmaceutical product or service

Set priorities for care

Step 1: Assess needs and identify drug therapy

Step 3: Monitor and review the care plan

Step 3: Develop a care plan

Step 3: Implement the care plan

Source: http://www.globalfundatm.org/contribute.html

The delivery of effective pharmaceutical care to patients requires pharmacists to practise in a new way that reflects their responsibility and accountability. However, due to limited resources, this step is not always possible and a systematic approach may need to be adopted to facilitate targeting of care. Such an approach is illustrated in Box 13.7.

Strand et al. used the term "pharmaceutical services" to represent all the services that pharmacists need to provide in order to resolve a patient's medicine therapy problems: these include medicine information and medicine distribution. Pharmaceutical care is a prospective patient-centred practice with a focus on identifying, resolving and preventing medicine therapy problems. Pharmacists require a high level of knowledge and skills to deliver pharmaceutical care, and an organizational structure to facilitate its delivery is required.

13.5.3 Medicines management

Continuous improvement of prescribing and the use of medicines remains one of the most crucial elements of healthcare development, aiming to ensure optimal health gain for patients and value for money for health care providers. The first challenge is to meet the changing needs of patients. The three priority areas are:

1. making sure that people can get medicines or pharmaceutical advice easily and, as far as possible, in a way, at a time and at a place of their choosing;
2. ensuring that they provide more support to patients in using medicines. Extra help for those who need it to get the best out of their medicines, help which will mean fewer people being ill because they are not using their medicines properly, and which will cut the amount of medicine which is simply wasted;
3. giving patients the confidence that they are getting good advice when they consult a pharmacist.

In Great Britain, a document entitled "Pharmacy in the Future" sets out the requirement for additional, structured professional support, to be provided by pharmacists, to optimise prescribing and to provide extra help, for those who need it, to get the best out of their medicines, plus mechanisms to reduce waste. It also states the aim of ensuring that patients get their medicines and have access to high quality pharmaceutical care conveniently, in a way and at a time and place of their choosing.

The intention is to improve and extend the range of pharmacy services available to patients, including identification of individuals' pharmaceutical needs, development of partnerships in medicines taking, co-ordination of repeat prescribing and dispensing processes, plus targeted treatment review and follow-up. This may also provide a model for the future of pharmacy elsewhere.

13.6 Conclusion

Science and technology offer both opportunities and threats in relation to pharmaceuticals in international health. The new sciences of pharmacogenetics and pharmacogenomics promise much in the way of new approaches to the prevention of illness and the treatment of disease. At the same time their cost threatens to increase the gulf in the levels of access between the peoples of developed and developing countries. For many, continued reliance on traditional remedies seems set to continue for some time: medicines available over the internet might as well be available on the moon.

If the health benefits of new technology and science are to reach populations, their application will need to be facilitated through the health workforce. Yet there remain enormous variations in the numbers of adequately trained health professionals available. However, as we have seen in this chapter, a number of strategies are now in place to ensure the rational use of existing medicines, and the introduction of new ones, through appropriate training and development.

But the successful application of scientific knowledge, the effective use of technological advances, and the maintenance of an adequately trained health workforce, all require appropriate policies at both national and international levels. We now turn to the wider policy issues in our final chapter.

Further reading

Bamford, C. (2003) "Genetics and genomics", *Health Futures* 1–10.

Cipolle, R.J., Strand, L.M. and Morley, P.C. (1998) *Pharmaceutical Care Practice* New York: McGraw Hill.

Javalgi, R. and Ramsey, R. (2001) "Strategic issues of e-commerce as an alternative global distribution system". *International Marketing Review* 18 (4): 376–391.

Murillo, L. (2001) "Supply chain management and the international dissemination of e-commerce", *Industrial Management and Data Systems* 101 (7): 370–377.

Official World List of Pharmacy Schools (WLPS) website. http//www.cf.ac.uk/phrmy/WWW-WSP/SoPListHomePage.html

Radford, T. (2001) *Evolution and Revolution: How Genomics will Change Health Care*. London: Association of the British Pharmaceutical Industry.

Richie, L. (2004) "E-commerce: Enabling efficient procurement in hospital pharmacy", *Pharmacy Management* 20 (1): 9–13.

Velasquez, G. and Boulet, P. (1999) "Essential drugs in the new international economic environment", *Bulletin of the World Health Organization* 77 (3), 288–292.

Chapter 14
Policy Initiatives and their Implications

Stuart Anderson and Rob Summers

Box 14.1: Learning objectives for chapter 14

By the end of this chapter you should be able to:

· Describe three product-based global public-private partnerships that concern medicines.
· Describe three product-development-based global public-private partnerships that concern medicines.
· Describe three systems/issues-based global public-private partnerships that concern medicines.
· List the objectives of the Global Alliance for Vaccines and Immunisation (GAVI).
· List the objectives of the Global Fund to fight AIDS, tuberculosis and malaria.
· Describe the WHO pre-qualification project.

14.1   Introduction

In this chapter we conclude our discussion of the management of pharmaceuticals in international health by exploring some of the directions policy is taking.

We begin by looking at some of the issues surrounding the development of public-private partnerships. This approach has been an area of rapid development in recent years and looks set to continue. As part of it we explore the increasing role of philanthropy in the provision of medicines in international health. We consider such initiatives as the Global Fund to fight AIDS, tuberculosis and malaria, and the Global Drug Facility for tuberculosis drugs.

We move on to a consideration of the global procurement of medicines, and the impact of initiatives such as the WHO pre-qualification project; and we return to the issue that has dominated discussion in the field of medicines in international health, i.e. medicines for AIDS. We consider where we are now in relation to the TRIPS agreement, post Doha and post "a dollar a day", and review the implications for both the industry and governments.

We conclude by exploring the prospects for increased cooperation between sectors, agencies and governments and the targeting of diseases or issues, and their potential for improving access to pharmaceuticals. Access to medicines is part of a much bigger problem around access to healthcare in general, to education and to everything else to which the world's people have every right, but where massive inequity still remains.

14.2   The role of public-private partnerships

We return now to some of the principal policy actors in the management of pharmaceuticals in international health. We have already noted that pharmaceutical companies are increasingly participating in, and indeed initiating, public-private partnerships. These partnerships are now bringing considerable additional resources into international public health and have the potential to benefit substantial populations. Benefits for the donor companies include the improvement of the corporate image, international recognition and influence, access to new markets and direct financial benefits, such as tax breaks.

Such partnerships are, however, not without their problems, which may involve real or perceived conflicts of interest. Unlike UN or WHO projects, they may be advised by so-called expert groups formed ad hoc based on personal contacts, with an inequitable representation of interests and competencies. Accountability is often directly to the donors and only indirectly to the public sector partner or recipient governments. While the partnerships benefit from substantial donations either financial or in kind, they also tie up public funds for operational expenses, which may cause shortages in other priority areas. The public funds involved may substantially exceed those provided by the private sector partner.

Lessons learnt from past global public-private partnerships have been summarised by Buse and Walt (see further reading), indicating that the following points are important for maximum effectiveness:

· Clearly specified, realistic and shared goals and reciprocal rights and obligations;
· Clearly delineated and agreed roles and responsibilities;
· Distinct benefits for all parties;
· The perception of transparency;
· Active maintenance of the partnership;
· Equality of participation;
· Meeting agreed obligations;
· Balance of power, based on beneficence, non-maleficence, autonomy and equity.

In a way, these experiences at the global level mirror those with public/private collaborations at the local level (see chapter 6, Box 6.7). Buse and Walt have developed a conceptual framework for understanding the different forms of global public-private partnership in the health sector, consisting of three types of partnership. We now use this framework to consider the present and future role of public-private partnerships in the management of pharmaceuticals in international health.

14.2.1   Product-based partnerships

Product-based partnerships consist mainly of medicine donation programmes. However, this category also includes those that undertake the bulk purchase of products for use in public sector programmes in low income countries. These include female condoms and AIDS medication. Selected examples of product-based partnerships are given in Box 14.2.

Medicine donation programmes are a favoured option when existing medicines are found to be effective in a disease or condition for which limited financial resources are available. This situation is usually due to lack of willingness or ability to pay. This type of partnership is usually initiated by the pharmaceutical industry. They seek a partnership with the multi-lateral sector in order to reduce costs and increase the chances of the medicine actually reaching those who need it but cannot afford it.

14.2.2   Product-development partnerships

Product-development partnerships differ from product-based partnerships in a number of important ways.

· They are not targeted at specific countries.
· They are generally initiated by the public sector.
· They are based on market failure to produce medicines which are required for specific conditions.

A number of pharmaceutical products that have been the subject of product-development partnerships are illustrated in Box 14.3. They include a number of medicines for malaria. Most of the medicines in this group are perceived by the public sector as worthy of societal investment. However, market mechanisms alone fail to allocate resources to their discovery and development, since the industry considers that the potential returns do not justify the opportunity cost of investment.

A notable feature of product-development partnerships is the retention of the intellectual property rights by the private sector partner, so as to retain leverage over eventual product pricing. Pharmaceutical companies may find product-development partnerships an attractive option:

· to mobilize a subsidy for research,
· to obtain assistance in carrying out clinical trials,
· to pursue their own longer term interests,
· to achieve at least a modest financial return,
· to seek proximity or involvement in standard setting and regulatory control, or
· to portray themselves in a favourable light in order to gain entry into emerging medicine markets.

Box 14.2: Examples of product-based public-private partnerships

---

Mectizan (Ivermectin) Donation Programme (from 1987)

Partners:	Merck and Co; Task Force on Child Survival and Development; WHO; World Bank; National authorities and NGOs
Goal:	To eliminate river blindness by treating everyone who needs it with Mectizan
Scope:	Medicine donated until no longer required; all 34 endemic countries have at some time been provided with free Mectizan; the cumulative value of the donations made is estimated at US $ 500 million; an additional US $ 0.2 million is spent on shipping plus the cost of the Mectizan Expert Committee and its Secretariat

---

Malarone (Proguanil and Atovaquone) Donation Programme (from 1996)

Partners:	Glaxo Wellcome; Task Force on Child Survival and Development; Medical Research Council, England; National Institutes of Health, USA; Centers for Diseases Control, Atlanta, USA; WHO; World Bank; Wellcome Trust; National authorities
Goal:	To help combat drug-resistant malaria in endemic countries where cost often limits access to new medicines
Scope:	Up to one million free doses per year globally through a targeted donation programme: pilot donations in Kenya and Uganda

---

Albendazole Donation Programme (from 1998)

Partners:	WHO/Division of Control of Tropical Diseases; SmithKline Beecham; Global Programme to Eliminate Filariasis; National authorities and NGOs
Goal:	To accelerate the effort to eliminate lymphatic filariasis
Scope:	Donation of albendazole to governments and other service providers until elephantiasis is eliminated: the value of the donation of up to six billion doses over twenty years is over US $ 1 billion

---

Source: Buse, K. and Walt, G. (2000) *Bulletin of the World Health Organization* 78 (5): 699–708.

14.2.3   Systems/services-based partnerships

The third group of partnerships is rather more eclectic. It includes some, such as the Malaria Vaccine Initiative, that arose in response to market failure, and others have been established to complement the efforts of governments, such as the Secure the Future partnership. Others have been established in order to tap into non-medical private resources for disease control, such as the World Alliance for Research and Control of Communicable Diseases. Examples of systems/services-based partnerships are listed in Box 14.4.

Some of these partnerships are characterised by the large number of partners and funders involved. The international AIDS vaccine initiative, for example, has

---

Box 14.3: Examples of product-development based public-private partnerships

---

Medicines for Malaria Venture (MMV) (from 1998)

Partners: Association of British Pharmaceutical Industries; International Federation of Pharmaceutical Manufacturers Associations; Wellcome Trust; Rockefeller Foundation; WHO/RBM/TDR; World Bank; Global Forum for Health Research; DFID; Swiss Development Cooperation (SDC); Glaxo Wellcome; Hoffman-La Roche

Goal: To support the discovery, development and commercialisation of affordable medicines for malaria at the rate of one every five years through a public sector venture fund

Scope: Public contribution of up to US $ 30 million per year: private sector to provide gifts in kind worth up to US $ 2 million per year: MMV retains patents for discoveries, and will license out projects for commercialisation to private companies; royalties retained for financial sustainability

---

LAPDAP (Chlorproguanil and Dapsone) (from 1998)

Partners: SmithKline Beecham; WHO/Tropical Disease Research Programme; DFID

Goal: To make available an affordable combination anti-malarial tablet

Scope: SmithKline Beecham, WHO and DFID to contribute one third of the development budget each

---

Malaria Vaccine Initiative (MVI) (from 1999)

Partners: Bill and Melinda Gates Foundation; PATH; private sector involvement through discovery and development partnership agreements

Goal: To accelerate the development of promising malaria vaccine candidates through identification and process development funding

Scope: Bill and Melinda Gates Foundation contributed founding grant of US $ 50 million

---

Source: Buse, K. and Walt, G. (2000) *Bulletin of the World Health Organization* 78 (5): 699–708.

---

five main partners and a large number of funders, including the World Bank and UNAIDS.

14.2.4 The future of public-private partnerships

We have noted in earlier chapters that the approach of the pharmaceutical industry has changed direction substantially during the course of the last thirty years. When the concept of essential drugs was first advocated in the 1970s the attitude of the pharmaceutical industry was one of extreme hostility. It did everything in its power to block progress, to minimise the implications of essential drugs for it, and to circumvent the policy in whatever ways it could.

Box 14.4: Examples of systems/services based public-private partnerships

---

Global Programme to Eliminate Filariasis (GPEF) (from 1998)

Partners: CDC; UNICEF, World Bank; WHO/CTD; DFID; SmithKline Beecham; Merck and Co; Arab Fund; Academia; Placer Dome Centre for International Health; international NGOs and national authorities

Goal: To eliminate lymphatic filariasis as a public health problem by the year 2020

Scope: Albendazole to be donated by SmithKline Beecham until elephantiasis eliminated (see Box 13.4); Mectizan to be donated by Merck and Co until onchocerciasis eliminated (see Box 13.4). All 73 endemic countries to be successively covered by programme

---

Bill and Melinda Gates Children's Vaccine Programme (CVP) (from 1998)

Partners: Bill and Melinda Gates Foundation; PATH; other partners with implementing role include UNICEF; WHO; World Bank; CVI; Ministries of Health; NGOs; academia; International Vaccine Institute; vaccine manufacturers

Goal: To reduce or eliminate existing time lag between developing and developed world in the introduction of new vaccines for children

Scope: Bill and Melinda Gates Foundation donated US $ 100 million as founding grant; industry contributes through donation of vaccines for model programmes, data for regulatory submissions, marketing information, and financial and market surveys: initial focus on three vaccines in 18 countries; ten year programme

---

Secure the Future (from 1999)

Partners: Bristol-Myers Squibb; UNAIDS; Harvard AIDS Institute; Medical Schools; National authorities

Goal: To improve HIV/AIDS research and community outreach in southern Africa

Scope: Bristol-Myers Squibb donated US $ 100 million for five year partnership; largest corporate donation for HIV/AIDS; covers Botswana, Lesotho, Namibia, South Africa and Swaziland

---

Source: Buse, K. and Walt, G. (2000) *Bulletin of the World Health Organization* 78 (5): 699–708.

This hostility continued in two main arenas; firstly, in the protracted negotiations that took place to develop the TRIPS agreement (which we have discussed in an earlier chapter); and secondly in the agreements that have now been reached concerning the manufacture and supply of medicines still under patent protection. The main group of such medicines is those used in the treatment of AIDS in developing countries. In these situations they can be sold at cost price, provided safeguards are in place to prevent their re-export to industrialised countries for sale at much higher prices.

The increasing participation in public-private partnerships of many of the global pharmaceutical companies, and their willingness to reach agreements leading to the greater availability of essential drugs to the people of low and middle income countries represents a significant shift in the stance of these companies. This shift has come about at least partly as a result of intense lobbying by NGOs, governments and international bodies such as WHO. But there are a number of benefits for the industry in becoming more conciliatory and cooperative.

· Involvement in global public-private partnerships has assisted companies to establish high level political contacts at the global and country levels.
· Global pharmaceutical companies are keen to establish their reputations as ethically oriented concerns.
· They hope that cooperation will result in good public relations exposure that will counter some of the bad publicity to which they have been subject in recent years and the poor practices which led to this type of publicity.

There have been many positive aspects to the emergence of public-private partnerships, but they have not been without their critics. These relate largely to matters of governance, representation, accountability and transparency. There are also many unanswered questions. What organizational forms and management arrangements represent best practice for governance, accountability and representation? And what factors contribute to partnership effectiveness on the ground? Research on these questions is required in order to develop guidelines, procedures and safeguards to harness the potential and minimise the risks associated with partnership arrangements.

Nevertheless, partnerships between the public and private sectors, including a wide range of commercial businesses, philanthropic foundations and both national and international organizations, offer some hope for tackling many of the issues addressed in this book. It is important that the goodwill and trust that has built up between them in recent years continues to grow and develop.

14.3   The Global Fund and other initiatives

Much has been achieved in pharmaceuticals since the introduction 25 years ago of the essential drugs and national drug policy concepts. Nearly 160 countries now have national essential drugs lists, whilst over 100 countries have national drug policies in place or under development. Most importantly of all, though, access to essential drugs increased in absolute numbers from 2.1 billion people in 1977 to 3.8 billion people in 1997.

The essential drugs concept is now widely accepted as a pragmatic approach to providing the best of modern, evidenced-based and cost-effective health care. It is even more valid today than it was 25 years ago when first introduced. Most development assistance organizations and NGOs, as well as public sector organizations, use essential drugs in their work.

Box 14.5: Objectives of GAVI

· Improve access to sustainable immunisation services.
· Expand the use of all existing safe and cost-effective vaccines where they address a pub-
  lic health problem and promote delivery of other appropriate interventions at immunisa-
  tion contacts.
· Accelerate the development and introduction of new vaccines and technologies.
· Accelerate research and development efforts for vaccines needed primarily in developing
  countries.
· Make immunisation coverage a centrepiece in international development efforts.
· Support the national and international accelerated disease control targets for vaccine preventable diseases.

Source: Vaccine Alliance (2003) http://www.vaccinealliance.org

---

14.3.1   The Global Forum for Health Research

But access to medicines is just one part of a much bigger issue concerned with
the availability of affordable and accessible healthcare. This issue is in turn one
part of the wider issue of access to clean water, adequate housing, a safe environment and education, and everything else to which every citizen of the world
has a right, but to which too many are denied access as a result of abject poverty. Those matters are beyond the scope of this book, but form the background
against which the management of pharmaceuticals in international health must
be viewed.

There is much that can be done. As we have seen, some of the major issues in the
management of pharmaceuticals in international health are now being addressed at
the global level. In recent years a number of global initiatives have been taken to
facilitate action. The Global Forum was established as a Foundation in 1998. This
organization uses two main strategies to correct the 10/90 gap: the development of
a methodology to identify priorities in health research, and the development of initiatives to bring together a wide range of partners to find solutions to priority health
problems. Further details about the Forum can be found on its website.

We conclude this discussion by highlighting two public-private partnerships that
represent possible models for the future development of global alliances in this area,
the Global Alliance for Vaccines and Immunization (GAVI) and the Global Fund.

14.3.2   The Global Alliance for Vaccines and Immunization

Recent developments have seen the emergence of large-scale initiatives involving
many partners with a focus on specific health problems. The GAVI is one such initiative. The objectives set for GAVI are given in Box 14.5.

> **Box 14.6: Objectives of the Global Fund to fight AIDS, tuberculosis and malaria**
>
> · To make available and leverage additional financial resources to combat AIDS, tuberculosis and malaria.
> · To base its work on programmes that reflect national ownership and respect country-led formulation and implementation processes.
> · To operate in a balanced manner in terms of different regions, diseases, and interventions, covering prevention, treatment, and care and support in dealing with the three diseases.
> · To establish a simplified, rapid innovative process with efficient and effective disbursement mechanisms, minimising transaction cost and operating in a transparent and accountable manner.
> · To support the substantial scaling up and increased coverage of proven and effective interventions, which strengthen systems for working within the health sector, across government departments, and with communities.
> · To focus on performance by linking resources to the achievement of clear, measurable and sustainable results.
>
> *Source*: Global Fund (2003) http://www.fundatm.org/contribute.html

GAVI was launched in January 2000 with an initial donation of US $ 750 million from the Bill and Melinda Gates Foundation. Following a decade of falling immunization coverage levels the aim of GAVI is to introduce new and underused vaccines, including hepatitis B and *Haemophilus influenzae* type B (Hib) vaccines, and to raise immunization coverage in the 74 poorest countries of the world.

14.3.3 The Global Fund to fight AIDS, tuberculosis and malaria

GAVI has been seen as a model for the new Global Fund to fight AIDS, tuberculosis and malaria. The objectives of the Global Fund are given in Box 14.6.

These objectives extend far beyond the provision of medicines and their rational use. They emphasise the need for simple, rapid and innovative procedures at all stages, and they focus firmly on performance by linking resources to the achievement of clear, measurable and sustainable results. There are fears that over-ambitious objectives set for the Fund, and excessive pressures placed on over-stretched Ministries of Health in individual countries, risk its failure in the longer term.

14.4 Global procurement of medicines

Global initiatives like these present enormous challenges for the procurement of medicines. The supply of sufficient quantities of medicines that are effective and of acceptable quality for the treatment of HIV/AIDS, malaria and tuberculosis has become a major concern at both international and country levels. Many international, regional and national organizations are involved in the procurement process.

14.4.1 WHO pre-qualification project

While WHO has had a formal pre-qualification system for vaccines for supply to UN organizations since 1989, no harmonised or uniform assessment system existed for pharmaceuticals until March 2001. Without such a system, organizations risk sourcing sub-standard, counterfeit and/or contaminated medicines. Inferior medicine quality can lead to drug resistance, product complaints and recalls, waste of financial resources and, most seriously, health risks to patients.

A pilot project on pre-qualification for procurement and sourcing of anti-malarial, anti-tuberculosis and HIV/AIDS medicines was launched by WHO in 2001. The project is managed by the Quality Assurance and Safety of Medicines Team of WHO's Department of Essential Drugs and Medicines Policies on behalf of WHO and other United Nations organizations such as UNAIDS, UNICEF, UNDP and UNFPA.

The prequalification project aims to facilitate access to medicines of acceptable quality through the assessment of compliance with WHO-recommended standards. The procedure for assessing the acceptability, in principle, of the medicines comprises a number of activities, which include:

· GMP inspection of the manufacturing sites.
· The evaluation of product data and information provided by the manufacturers and suppliers.

Only products and manufacturing sites that are found to meet the recommendations as stipulated in the WHO guidelines are published in the list of approved suppliers. This list can then be used to select appropriate suppliers for procurement of the medicines required by the global partnerships.

The prequalification product supports the AIDS Medicines and Diagnostics Service (AMDS), set up as part of the 3 by 5 strategy launched in December 2003, which aims to provide HIV treatment to 3 million patients by the year 2005.

14.4.2 Global Drug Facility for tuberculosis drugs

The Global Drug Facility, hosted by WHO and operated by the Stop TB Partnership, has supplied procurement support and medicines to 2.8 million TB patients in 65 countries since its launch in March 2001. In November 2003, the WHO and the pharmaceutical company Novartis signed an agreement, under which Novartis is to donate tuberculosis drugs to the Global Drug Facility. Treatment for a total of half a million patients will be provided free of charge over a five-year period to countries scaling up TB control with support from the Global Fund to fight AIDS, tuberculosis and malaria.

Novartis will supply special patient kits containing fixed-dose combination tablets in blister packs. The design improves patient compliance and reduces the risk of developing drug-resistant TB. The drugs will be supplied to programmes using the Directly Observed Treatment Short course (DOTS), the internationally recommended strategy for tuberculosis control. More than 10 million people have been successfully treated under DOTS since 1993, when WHO declared tuberculosis to be a global emergency.

14.5 AIDS drugs post Doha

The historic Doha Declaration, made at the November 2001 WTO Ministerial Conference, affirmed that the TRIPS agreement can and should be interpreted to promote public health and access to medicines "for all". Key flexibilities available in TRIPS include granting compulsory licensing and parallel imports of cheaper generic drugs. Using generic versions of HIV/AIDS medicines from India, for example, would allow the least developed countries to treat three times as many people than with patented brand-name versions. Prices of a patented triple cocktail of ARV were at first approximately US $ 10,000 per person per year.

Due to a combination of public pressure and generic offers from Indian firms, companies then cut prices to around US $ 900. In October 2003 the Clinton Foundation announced that it had brokered a deal with four generics companies to provide triple-drug antiretroviral therapy to governments in the developing world at a cost of less than US $ 140 per patient per year. The environment will change when the TRIPS agreement, which grants twenty-year product patent protection, is introduced in developing countries.

14.5.1 Supply and demand

Suppliers of medicines have to cover not only the cost of production, but also development costs. The key idea underpinning the TRIPS agreement is to provide an incentive for private parties to undertake R&D. It has been suggested to replace TRIPS by an R&D treaty, under which countries would have to devote a proportion of their GDP to R&D, but could carry out the latter in any way they chose, be it with public sector support or through the patent system. The treaty would emphasise public health priorities such as neglected diseases and vaccines, promote technology transfer and facilitate transparency of investment flows.

On the demand side, the main interest is to buy products from the cheapest provider, without regard to who holds patents. Practices akin to compulsory licensing are used extensively in some First World countries, including "government use" in the US, and "Crown Use" in the UK, where government officials can authorize third-party use of a patent without previous negotiation. Patent holders can then sue for compensation. Governments of developing countries, on the other hand, remain reluctant to take advantage from the TRIPS provisions for a number of reasons, which include direct political pressure from dominant countries, although their demand for cheap generic medicines is great and has a major impact on public health.

14.5.2 Implications for the industry

The pharmaceutical industry comprises both large multinational companies with their own R&D units, as well as large and small scale generic manufacturers. Implications of the TRIPS provisions for these segments of the industry differ.

From 2005 onwards, generic manufacturers will not be allowed to produce medicines patented within the previous ten years. In order to survive, generic manufacturers will need to develop their own research and development capabilities, or to collaborate with large pharmaceutical firms capable of these investments, bearing in mind that many multinational companies have their own generics divisions. Even when generic firms are granted permission to manufacture drugs for certain "emergency" conditions such as HIV/AIDS, new stipulations made at the Cancun convention in 2003 essentially demand that they do so on a humanitarian basis, and not for profit. Such stipulations make it very difficult for generics manufacturers to maintain viable operations.

Large multinational companies have been far more willing to cut their prices than to make concessions with regard to their patent rights. They can afford to lower their prices in the Third World market, but are likely to continue setting global priorities for research, development and pricing of medicines thanks to the patent protection which they enjoy.

14.5.3 Implications for governments

In order to benefit from the flexibility provided by the TRIPS agreement, most developing countries need to amend their own national legislation. Technology transfer is also needed to build production capacity where possible.

Reasons which have kept governments of developing countries from making good use of TRIPS flexibilities include continued pressure from multinationals and the United States government, fear of sanctions even where none are likely ("the chilling effect"), legal uncertainty, poor coordination amongst government ministries, the nature of technical assistance provided by international organizations such as the WIPO, which often emphasises countries' obligations under TRIPS rather than their rights, and fear of losing direct foreign investment.

Governments need to take advantage of the flexibility provided by the TRIPS agreement. Consultations have been held, and a manual has been made available to provide technical guidance for this purpose. Cooperation between governments, extending not only to information-sharing but also to production, technology-sharing, distribution, import and export of medicines, would strengthen these countries' position. (The TRIPS agreement is described in more detail in chapter 8.)

14.6   Conclusion

Constructive and multiple solutions are needed in order to reduce the inequities in access to medicines, and at the same time protect the incentives that are needed to ensure that vital research and development is continued. Despite an apparently bleak situation many steps have been taken in recent years to improve access to medicines and treatment. Increasingly, a diverse range of local and international approaches is being taken within an overall global strategy.

There are grounds for optimism that such coordinated and synergistic initiatives might make a positive contribution to ensuring that all people have access to quality medicines and treatment. The Global Fund and other initiatives offer an important opportunity to build sustainable health systems in order to protect the world's poor. But they can be only part of the solution. Philanthropy is an important additional measure, but it becomes part of the problem if states use it as an excuse to restrict further funding of their own.

It is within such a framework that the problems of lack of access to essential medicines stand the best chance of solution. It is vital that the opportunity is taken, and that the Fund does not become just another short-term, high-profile political goal, with easy access to its resources for organizations proficient in grant-writing.

Access to health care and pharmaceuticals is too important for governments to be given any opportunity to escape their responsibilities. The importance of the health care system within which pharmaceuticals are managed in international health cannot be overstated. Medicines are needed, obtained and consumed within the wider political, economic, social and cultural context within which people live their lives. Pharmaceutical supply is one thing, equitable and sustainable access to medicines is another.

We hope that this book has demonstrated that, although there are no easy solutions to improving access to affordable medicines of good quality by the people of low and middle-income countries, there are many things that can be done, by many actors in many different ways, to make a difference. We would like to think that the book has highlighted some of them. Finally, we hope that the book has helped give readers the skills, knowledge and confidence, and perhaps also the inspiration, to make their own contribution to this important and rewarding field of health care.

Further reading

Brugha, R., Starling, M. and Walt, G. (2002) "GAVI, the first steps: lessons for the Global Fund", *Lancet* 359: 435–358.

Buse, K. and Walt, G. (2000) "Global Public-Private Partnerships: Part II-What are the health issues for global governance?", *Bulletin of the World Health Orga-*

nization 78 (5): 699–708. http://www.who.int/docstore/bulletin/pdf/2000/
issue5/bu0241.pdf.

Correa, C. (2000) *Intellectual Property Rights, the WTO and Developing Countries: The TRIPS Agreement and Policy Options.* New York: Zed Books, Third
Work Network.

Global Drug Facility (2004) *An Initiative of the Global Partnership to Stop TB.*
http://www.stoptb.org/GDF/default.asp.

Global Fund (2004) http://www.theglobalfund.org/en/.

*Implementation of the Doha Declaration on the TRIPS Agreement and Public
Health: Technical Assistance – How to Get it Right* (2002) Conference Report.
28th March 2002, International Conference Centre of Geneva (CICG):
Médecins sans Frontières (MSF), Health Action International (HAI), Consumer
Project on Technology (CPT), Oxfam International.

Love, J. (2003) *From TRIPS to RIPS: A Better Trade Framework to Support Innovation in Medical Technologies.* Presented at the Workshop on Economic issues
related to access to HIV/AIDS care in developing countries, Université de la
Méditerranée, Marseille, France, May 27th, 2003. http://www.cptech.org/
slides/trips2rips.doc.

Quick, J. (2003) "Essential medicines twenty-five years on: closing the access gap",
Health Policy and Planning 18 (1): 1–3.

Quick, J. and Hogerzeil, H. (2003) "Ten best readings in ... essential medicines",
Health Policy and Planning 18 (1): 119–121.

US Bullying on Drug Patents: One Year after Doha (2002) Oxfam Briefing Paper
No. 33. http://www.oxfam.ca/campaigns/downloads/TRIPSbriefUSbullying2.
pdf.

Velasquez, G. and Boulet, P. (1999) "Essential drugs in the new international economic environment", *Bulletin of the World Health Organization* 77 (3): 288–
292.

WHO Pre-Qualification Scheme (2004) Geneva: World Health Organization. Pre-
qualification project website. http://mednet3.who.int/prequal/.

Widdus, R. (2001) "Public-private partnerships for health: their main targets, their
diversity, and their future directions", *Bulletin of the World Health Organization* 79 (8): 713–720.

Selected Websites and Webportals

The number of websites and webportals with information relevant to the management of pharmaceuticals in international health is now vast. We include here a small number that we think are of particular relevance, which are based on a more extensive list compiled by Reinhard Huss and Ingo Glueckler at the University of Heidelberg. Many contain information from a large number of categories, and the headings below are for guidance only. The MSH, UN and WHO websites are listed separately.

Antimicrobial resistance

· *Alliance for the Prudent Use of Antibiotics* (APUA)
Antibiotic discussion group for scientists and health professionals in fields of infectious diseases and microbiology. Disseminates scientific information on antibiotic resistance and promotes rational use of antibiotics.
Contact: anibal.sosa@tufts.edu

Drug bulletins

· *Arznei-Telegramm* (German Bulletin with English articles)
http://www.arznei-telegramm.de/
· *Boletin Farmacos* (Spanish Bulletin for subscribers in Latin America)
http://www.boletinfarmacos.org
· *Drug and Therapeutics Bulletin* (Independent evaluations of drugs and other treatments for British health professionals)
http://www.which.net/health/dtb/main.html
· *International Society of Drug Bulletins*
http://www.isdbweb.org/
· *La Revue Prescrire*
French Bulletin with English translation published as Prescrire International
http://www.esculape.com/prescrire
· *The Medical Letter on Drugs and Therapeutics*
Critical appraisals of new drugs and comparative reviews of older drugs
http://www.medicalletter.com

Educational institutions

- *Boston University Medical Center (BUMC)*
  Richard Laing's webpage at BUMC provides direct access to many of the websites listed here, plus a wide range of training material.
  http://dcc2.bumc.bu.edu/richardl/IH820/resources.htm
- *Liverpool School of Tropical Medicine*
  http://www.liv.ac.uk/lstm
- *London School of Hygiene and Tropical Medicine*
  http://www.lshtm.ac.uk
- *Medical University of Southern Africa*
  http://www.medunsa.ac.za
- *Swiss Tropical Institute*
  http://www.sti.ch
- *tropEd (The European Network for Education in International Health)*
  http://www.troped.org
- *University of Heidelberg*
  http://www.hyg.uni-heidelberg.de/ithoeg

Electronic publications

- *British Medical Journal*
  http://www.bmj.com
- *Geneva Medical Foundation* (Information about open access journals)
  http://www.gfmer.ch/Medical_journals/General_medicine.htm
- *International Network for the Availability of Scientific Publications (INASP)*
  Provides health links to selected websites that are of special interest to health professionals, health service users and NGOs. http://www.inasp.info/
- *Lancet*
  http://www.thelancet.com
- *Oxford University Press*
  Free or greatly discounted electronic access to a large number of professional journals available to scholars from developing nations
  http://www3.oup.co.uk/jnls/devel/

Essential and generic drug suppliers

- *ECHO* (United Kingdom): http://www.echohealth.org.uk
- *Generic Drugs* (Brazil): http://www.medicamentogenerico.org.br/
- *IDA* (The Netherlands): http://www.ida.nl
- *MEDEOR* (Germany): http://www.medeor.org and http://order.medeor.org
- *MISSIONPHARM* (Denmark): http://www.missionpharma.com
- *UNIPAC* (Unicef): http://www.supply.unicef.dk

European institutions

- *EU Directorate General for Development*
  http://europa.eu.int/comm/dgs/development/organisation/mission_en.htm
- *EU Sixth Framework*
  http://europa.eu.int/comm/research/fp6/index_en.html
- *EuropeAid*
  http://europa.eu.int/comm/europeaid/general/mission_en.htm

Evidence based medicine

- *Centre for Evidence-Based Medicine* (Oxford, UK)
  http://www.cebm.net/study_designs.asp
- *Clinical Evidence*
  British Medical Journal publication. http://www.clinicalevidence.com
- *Cochrane Collaboration*
  http://www.cochrane.org
- *Cochrane Library*
  Abstracts of Cochrane Reviews available without charge
  http://www.update-software.com/Cochrane/default.HTM
- *Data Base of Abstracts of Review of Effectiveness* (DARE)
  http://www.update-software.com/ccweb/cochrane/cdsr.htm

Formularies

- *Belgian National Formulary* (French and Dutch)
  http://www.cbip.be
- *British National Formulary*
  http://www.bnf.org
- *Mercy Ships Formulary*
  Formulary based on WHO model produced by NGO. Reference book containing information on dosage, availability and cautions. For use by medical personnel at bedside. http://www.mercyships.org and http://www.drugref.org
- *South African Medicines Formulary*
  http://www.uct.ac.za/depts/mmi/jmoodie/editor.html

Health information resources

- *HealthNet*
  Administered by SATELLIFE, a non-profit organization based in Massachusetts. HealthNet is a global communications network, which links healthcare workers around the world via e-mail. HealthNet can be used to link libraries,

access information for individuals, carry informational updates, and create discussion groups. http://www.healthnet.org
· *HealthNet News*
Selects relevant content from over 45 journals and provides either full text information or abstract in one of three e-mailed newsletters (also HealthNet News AIDS and HealthNet News Community Health). Available free of charge to subscribers in low resource countries.
Email requests to: hnet@healthnet.org
· *Id21healthnews*
The id21 online collection contains hundreds of policy-relevant research digests on global development issues. To subscribe to id21HealthNews send email request to: lyris@lyris.ids.ac.uk. Website at: http://www.id21.org/
· *Health Systems Information*
Eldis Health Systems Resource Guide. A resource centre sponsored by DFID. Topics include poverty, priority diseases, health policy, access to medicines, and health service delivery. http://www.eldis.org/healthsystems/

Health professions

· *Department of Health: The NHS Plan, Pharmacy in the Future*
http://www.doh.gov.uk/pharmacyfuture/index.htm
· Dr Libby Roughead et al. *The Value of Pharmacist Professional Services in the Community Setting: A systematic review of the literature 1990–2002.* Available at: http://www.guild.org.au/public/researchdocs/reportvalueservices.pdf

HIV/AIDS drugs

· *ACTIS: AIDSDRUGS/AIDSTRIALS HIV/AIDS*
Clinical trials information http://www.actis.org/
· *AIDS Drug Interactions*
http://www.hiv-druginteractions.org
· *AIDSLINE*
http://gateway.nlm.nih.gov
· *AIDSMAP/NAM*
NAM produces extensive information on treatments, both in book form and as a searchable database. http://www.aidsmap.com/main/hatip.asp
· *Anti-retroviral therapy guidelines*
Follow the links to HIV treatment guidelines at: http://www.cdc.gov
· *Clinical Guidelines for HIV/AIDS in South Africa*
http://www.aidforaids.co.za/ClinicalGuidelines/introduction.html
· *Delivering HIV treatment*
Discussion summary. http://www.id21.org/hiv/report.html

· *Sources and Prices*
Selected drugs and diagnostics for people living with HIV/AIDS can be found at
http://www.unicef.org/supply
http://www.unaids.org
http://www.who.int/medicines
http://www.accessmed-msf.org

Intellectual property rights

· *Report of the Commission on Intellectual Property Rights (DFID)*
Integrating Intellectual Property Rights and Development Policy
http://www.iprcommission.org/
· *Information about Trade-Related aspects of Intellectual Property Rights
(TRIPS)* for policymakers, NGOs and others by the British NGO Panos
http://www.panos.org.uk/
· *Consumer Project on Technology*
http://www.cptech.org

Management Sciences for Health (MSH)

Private, non-profit educational and scientific organization working to close the
gap between knowledge and action in public health. http://www.msh.org

· *International Network for Rational Use of Drugs (INRUD)*
Bulletin published twice yearly (Hardcopy and electronic)
http://www.msh.org/inrud/
· *Strategies for Enhancing Access to Medicines (SEAM)*
http://www.msh.org/seam
· *International Drug Price Indicator Guide*
http://www.msh.org/inrud/activities.html
linked to Drug Prices on: http://www.who.int/medicines

Medicine information resources

· *American Hospital Formulary Service (AHFS)*
Aims to provide an evidence-based foundation for safe and effective drug ther-
apy. Provided by the American Society of Healthcare Pharmacists.
http://www.ashp.org and http://www.ashpdruginformation.com
· *Assessing Medicines Information*
Grassian, Esther. *Thinking Critically about World Wide Web resources*. UCLA
College Library.
http://www.library.ucla.edu/libraries/college/help/critical/index.htm

- *DIRLINE (Directory of Information Resources Online)*
  The National Library of Medicine's online database containing a wide variety of information resources including organizations, research resources, projects, and databases concerned with health and biomedicine. DIRLINE contains approximately 10,000 records and focuses primarily on health and biomedicine, although it also provides limited coverage of some other special interests. http://dirline.nlm.nih.gov
- *DrugInfoZone*
  A medicines information knowledge base designed for healthcare professionals in the UK's National Health Service (NHS). Aims to promote safe, effective and efficient use of medicines within the NHS. Content is independent, unbiased and evidence based. http://www.druginfozone.org/
- *eFacts*
  A list of browser-based products that includes Drug Facts and Comparisons, Drug Interaction Facts, The Review of Natural Products, Cancer Chemotherapy, Nonprescription Drug Therapy, Off-Label Drug Facts, Med Fact, Manufacturer Index, Drug Identifier. http://www.factsandcomparisons.com
- *eMC*
  Provides Data Sheets and Summaries of Product Characteristics (SPCs) for 2,500 medicines licensed in the UK. http://emc.vhn.net/
- *International Network of Drug Information Centres (INDICES)*
  A global network of drug information centres that advocate rational drug therapy. INDICES helps with specific drug information enquiries. http://www.essentialdrugs.org/indices/about.php
- *Keeping up to date*
  De Vries, T.P, Henning, R.H, Hogerzeil, H.V. and Fresle, D.A. (1994) *Guide to Good Prescribing*. Chapter 12: "How to keep up-to-date about drugs". WHO/DAP/95.1.
  http://www.med.rug.nl/pharma/who-cc/ggp/chapterc/page01.htm
- *Medline Plus Drug Information*
  A guide to more than 9,000 prescription and over-the-counter medicines provided by the United States Pharmacopoeia (USP) in the USP DI® Advice for the Patient®. http://www.nlm.nih.gov/medlineplus/druginformation.html
- *Merck Manual of Diagnosis and Therapy*
  Provides clinical and prescribing information for healthcare professionals. Includes a keyword search engine. http://www.merck.com/pubs/mmanual/
- *National Medicines Management Services Programme*
  A collaborative programme in the UK National Health Service.
  http://www.doh.gov.uk/pharmacyfuture/medicinesmanagement1.htm
- *PDR Online*
  Provides access to the online version of Physicians' Desk Reference, PDR for herbal medicines and PDR multi-drug interactions. The content has a US focus. http://physician.pdr.net/
- *PharmaInfoNet Glossary*
  A guide to pharmacological terms. http://bovis.gyuvet.ch/3dict/344farm1.htm

- *Pharma-Lexicon*
  A dictionary of pharmaceutical medicine. Database with over 28,000 medical abbreviations and acronyms and over 30 million scientific articles. http://www.pharma-lexicon.com
- *PharmWeb*
  Provides links to pharmaceutical information on the internet including companies, regulatory bodies, societies and publishers. *Pharmsearch* is a specialized search engine designed to focus on pharmaceutical and health-related information. http://www.pharmweb.net/
- *Virtual Library: Pharmacy*
  A gateway to drug-related databases, journals, companies, toxicology, conferences and organizations. Includes a key word search engine. http://www.pharmacy.org/
- *Virtual Pharmacy Center*
  A gateway to pharmacy, pharmacology, clinical pharmacology and toxicology, related conferences and organizations. Part of Martindale's Health Science Guide. http://www.martindalecenter.com/Pharmacy.html

Medicines promotion

- *The Drug Promotion Database*
  A comprehensive compilation of almost 2,200 entries about all aspects of drug promotion, is a joint project of the WHO and Health Action International. http://www.drugpromo.info/

Medicines utilization data

- *Denmark* medicines consumption data available at http://www.laegemiddelstyrelsen.dk and http://www.dkma.dk
- *Norway* medicines consumption data for 1998–2002 available at http://www.legemiddelforbruk.no/English/
- *United Kingdom* prescription statistics available at http://www.doh.gov.uk/prescriptionstatistics/index.htm

Non-governmental organizations

- *BUKO*
  Pharma Kampagne, Germany
  http://www.epo.de/bukopharma.html
- *BUMC*
  Promoting Rational Drug Use, MS Word and MS PowerPoint Course Materials. http://dcc2.bumc.bu.edu/prdu/Word_Powerpoint_Files_TOC.html

- *DNDI*
  Drugs for Neglected Diseases Initiative. http://www.dndi.org
- *ECHO*
  http://europa.eu.int/comm/echo/index_en.htm
- *GPHF*
  German Pharma Health Fund, The Minilab. http://www.gphf.org
- *Health Action International (HAI)*
  http://www.haiweb.org/
- *Médecins Sans Frontières (MSF)*
  http://www.accessmed-msf.org
- *OXFAM*
  http://www.oxfam.org.uk
- *Sphere Project*
  http://www.sphereproject.org

Patient information about medicines

- *Practice Guidelines for the Labelling and Packaging of Medicines*
  Guidelines produced by the UK Medicines Control Agency (MCA).
  http://www.mca.gov.uk
- *Communication Research Institute of Australia (CRIA)*
  Specialises in developing principles for writing useable information on medicines. http://www.communication.org.au

Pharmaceutical industry

- *International Federation of Pharmaceutical Manufacturers Associations*
  http://www.ifpma.org
- *Pharmaceutical Research and Manufacturers of America*
  http://www.phrma.org/
- *Research-based Pharmaceutical Industry in Germany*
  http://www.vfa.de
- *Association of the British Pharmaceutical Industry*
  http://www.abpi.org.uk
- *International Conference on Harmonisation of Technical Requirements for Registration of Pharmaceuticals for Human Use (ICH)*
  Collaboration between Pharmaceutical Industry and Drug regulatory authorities in USA, Japan and EU. http://www.ich.org

Pharmacopoeias

- *British Pharmacopoeia*
  http://www.britpharm.com or http://www.pharmacopoeia.org.uk

- *United States Pharmacopoeia*
  Standards for over 3,700 medicines, dietary supplements and dosage forms
  http://www.usp.org/
- *International Pharmacopoeia*
  Quality specifications for pharmaceutical substances and tablets
  http://www.who.int/medicines/library/pharmacopoeia/pharmacop-content.
  shtml

Pharmacovigilance

- *Antidepressants*
  Website exploring threats to public safety and academic freedom surrounding
  the SSRI group of antidepressant drugs: http://www.healyprozac.com/
- *Rapid Alert*
  WHO issues a "rapid alert" whenever a serious problem in the safety of any
  medicinal product arises. The alerts are posted on the WHO EDM website.
  http://www.who.int/medicines/organization/qsm/activities/drugsafety/orgqsm
  alerts.shtm

Pricing of medicines

- *Australian Pharmaceutical Benefits Scheme*
  http://www.health.gov.au/pbs/
- *French Generic Drug Prices*
  Guide des Equivalents thérapeutiques http://www.ameli.fr.
- *WHO Drug Prices*
  http://www.who.int/medicines/organization/par/ipc/drugpriceinfo.shtml or
  http://erc.msh.org/mainpage.cfm?file=4.0.cfm&id=12&temptitle=Refer-
  ence&module=DMP&language=English

Public interest and consumer organizations

- *Choice* (Australia)
  http://www.choice.com.au
- *Consumer Institute for Medicines and Health* (Sweden)
  http://www.kilen.org
- *Global Treatment Access Campaign*
  http://www.globaltreatmentaccess.org/
- *Health Action International*
  http://www.haiweb.org/

Public-private partnerships

- *Global Alliance for TB Drug Development*
  http://www.tballiance.org
- *Global Drug Facility (GDF)*
  http://www.tballiance.org/4_3_3_GlobalDrugFacility.asp
- *Global Fund to fight Tuberculosis, AIDS and Malaria*
  http://www1.theglobalfund.org/en/

Quality of medicines

- *United States Pharmacopoeia Drug Quality and Information (USP DQI)*
  Funded by USAID and works in collaboration with WHO/PAHO in all aspects
  of drug quality. Focuses on TB, HIV AIDS and malaria. Deals with drug quality assurance at the central and provincial levels, as well as with manufacturers, DRA, Procurement and MOH. http://www.uspdqi.org

Rational use of medicines

- *Australian Prescribing Service*
  Set up to promote rational drug use. http://www.nps.org.au
- *International Conference on Improving Use of Medicines (ICIUM) 2004*
  http://www.who.int/medicines/organization/par/icium/icium.shtml
- *International Conference on Improving Use of Medicines (ICIUM) 1997*
  http://www.who.int/dap-icium/summary.html
- Additional training material available at
  http://dcc2.bumc.bu.edu/prdu/default.html

Regulatory authorities

Addresses of National Drug Regulatory Authorities, Pharmacovigilance centers
and other useful information can be found on the website of the WHO Collaborating Center for International Drug Monitoring. http://www.who-umc.org

- *European Medicines Evaluation Agency (EMEA)*
  http://www.emea.eu.int/
- *Food and Drug Administration (FDA) USA*
  http://www.fda.gov/cder/drug/default.htm
- *Medicines and Healthcare Products Regulatory Agency (UK)*
  http://www.mca.gov.uk/

Traditional medicines

- *French Society of Ethnopharmacology*
  http://www.ethnopharmacologia.org
- WHO Fact Sheet on *Traditional and Alternative Medicine*, 2003.
  http://www.who.int
- *WHO/TRM*, Regulatory situation of Herbal Medicines, WHO/TRM/98.1.
  http://www.who.int

Treatment guidelines

- *Australian Guidelines*
  http://www.tg.com.au/
- *Colombian Guidelines* (in Spanish):
  http://ascofame.org.com
- *Finnish Guidelines*
  http://www.ebm-guidelines.com
- *Guidelines International Network*
  http://www.guidelines-international.net
- *Links to Guidelines mainly for industrialised countries*
  http://www.sheffield.ac.uk/~scharr/ir/netting/
- *Scottish Intercollegiate Guidelines Network (SIGN)*
  http://www.sign.ac.uk/guidelines/html
- *South African Department of Health Fact Sheets and Guidelines*
  http://www.doh.gov.za/docs/factsheets

United Nations

- *United Nations* website: http://www.un.org/
- *UNAIDS*
  http://www.unaids.org/publications/documents/health/access/patsit.doc

World Health Organization (WHO)

- WHO website: http://www.who.int
- Select Health topics A–Z, Select Medicines & Vaccines for
  Essential Drugs and Medicines Policy (EDM) website:
  http://www.who.int/medicines/
- Provides access to the following topics:
  Access to medicines; Drug regulation; Essential drugs and medicines; Essential
  drugs lists; Drug prices; Essential Drugs Monitor; Financing mechanisms of
  medicines; Global Alliance for Vaccines and Immunization; Good Manufac-

turing Practice (GMP); International Non-proprietary Names (INN); Medicinal plants; Narcotic and psychotropic drugs; Policy, Access & Rational Use of
Drugs; Safety and standards of medicine; Vaccines & immunization

· *WHO and PAHO Health Library for Disasters 2001*
  http://who.int/eha/disasters/
· For access to HIV/AIDS Drugs and Diagnostics of Acceptable Quality; for the
  Prequalification of HIV/AIDS drugs; Collaboration of UNAIDS; WHO,
  UNICEF, WORLD BANK:
  http://www.unfoundation.org/unwire/util/display_stories.asp?objid=24859
· *WHOLIS (Library Catalogue and WHO Full Text Articles)*
  http://dosei.who.int/uhtbin/cgisirsi/Wed+Nov+21+23:21:20+MET+2001/0/49
· *What's new (Publications)*
  http://www.who.int/medicines/information/infnews2.shtml
· *Guide to Good Prescribing*
  http://www.med.rug.nl/pharma/who-cc/ggp/homepage.htm
· *Training Manual available at:* http://www.who.int/medicines/
· *Drug Prices:*
  http://www.who.int/medicines/organization/par/ipc/drugpriceinfo.shtml and
  http://erc.msh.org/mainpage.cfm?file=4.0.cfm&id=12&temptitle=Refer-
  ence&module=DMP&language=English
· *Essential Drugs List*
  Organised by therapeutic category and drug name.
  http://www.who.int/medicines/organization/par/edl/eml.shtml.
· *WHO Formulary Online*
  The WHO Formulary is now searchable online at:
  http://mednet3.who.int/mf/modelFormulary.asp or http://mednet3.who.int/mf/
  It can be downloaded as an Adobe PDF file (2.19 MBytes)
· *Reporting detected counterfeit drugs*
  http://www.who.int/medicines/organization/qsm/activities/qualityassurance/co
  unterfeit/reporting_cd.shtml

World Trade Organization (WTO)

· *WTO, WHO secretariats workshop on affordable drugs.*
  http://www.wto.org/english/tratop_e/trips_e/tn_hosbjor_e.htm

Index